NATURAL ENEMIES HANDBOOK

NATURAL ENEMIES HANDBOOK

The Illustrated Guide to Biological Pest Control

Mary Louise Flint
Department of Entomology
and
University of California Statewide Integrated Pest Management Project
University of California, Davis

Steve H. Dreistadt
University of California Statewide Integrated Pest Management Project
University of California, Davis

Photographs by Jack Kelly Clark

UNIVERSITY OF CALIFORNIA
STATEWIDE INTEGRATED PEST MANAGEMENT PROJECT

UC DIVISION OF AGRICULTURE AND NATURAL RESOURCES
and the
UNIVERSITY OF CALIFORNIA PRESS
Berkeley   Los Angeles   London

PRECAUTIONS FOR USING PESTICIDES

Pesticides are poisonous and must be used with caution. READ THE LABEL BEFORE OPENING A PESTICIDE CONTAINER. Follow all label precautions and directions, including requirements for protective equipment. Use a pesticide only against pests specified on the label or in published University of California recommendations. Apply pesticides at the rates specified on the label or at lower rates if suggested in this publication. In California, all agricultural uses of pesticides must be reported, including use in parks, golf courses, roadsides, cemeteries, and schoolyards. Contact your county agricultural commissioner for further details. Laws, regulations, and information concerning pesticides change frequently, so be sure the publication you are using is up-to-date.

Legal Responsibility. The user is legally responsible for any damage due to misuse of pesticides. Responsibility extends to effects caused by drift, runoff, or residues.

Transportation. Do not ship or carry pesticides together with food or feed in a way that allows contamination of the edible items. Never transport pesticides in a closed passenger vehicle or in a closed cab.

Storage. Keep pesticides in original containers until used. Store them in a locked cabinet, building, or fenced area where they are not accessible to children, unauthorized persons, pets, or livestock. DO NOT store pesticides with foods, feed, fertilizers, or other materials that may become contaminated by the pesticides.

Container Disposal. Dispose of empty containers carefully. Never reuse them. Make sure empty containers are not accessible to children or animals. Never dispose of containers where they may contaminate water supplies or natural waterways. Consult your county agricultural commissioner for correct procedures for handling and disposal of large quantities of empty containers.

Protection of Non-Pest Animals and Plants. Many pesticides are toxic to useful or desirable animals, including honey bees, natural enemies, fish, domestic animals, and birds. Certain rodenticides may pose a special hazard to animals that eat poisoned rodents. Crops and other plants may also be damaged by misapplied pesticides. Take precautions to protect non-pest species from direct exposure to pesticides and from contamination due to drift, runoff, or residues.

Posting Treated Fields. For some materials, re-entry intervals are established to protect field workers. Keep workers out of the field for the required time after application and, when required by regulations, post the treated areas with signs indicating the safe re-entry date.

Harvest Intervals. Some materials or rates cannot be used in certain crops within a specified time before harvest. Follow pesticide label instructions and allow the required time between application and harvest.

Permit Requirements. Many pesticides require a permit from the county agricultural commissioner before possession or use.

Processed Crops. Some processors will not accept a crop treated with certain chemicals. If your crop is going to a processor, be sure to check with the processor before applying a pesticide.

Plant Injury. Certain chemicals may cause injury to plants (phytotoxicity) under certain conditions. Always consult the label for limitations. Before applying any pesticide, take into account the stage of plant development, the soil type and condition, the temperature, moisture, and wind. Injury may also result from the use of incompatible materials.

Personal Safety. Follow label directions carefully. Avoid splashing, spilling, leaks, spray drift, and contamination of clothing. NEVER eat, smoke, drink, or chew while using pesticides. Provide for emergency medical care IN ADVANCE as required by regulation.

ISBN 1-879906-37-6 (hardbound)
ISBN 1-879906-41-4 (paperbound)
ISBN 0-520-21801-9 (University of California Press edition)
Library of Congress Catalog Card No. 97-062438

© 1998 by the Regents of the University of California
Division of Agriculture and Natural Resources

Printed in Canada.

15.5m–pr–5/98–NS/IPM

Dedication

THIS BOOK IS DEDICATED to the memory of Ken Hagen, Robert van den Bosch, and Carl Huffaker. It was their encouragement and enthusiasm for biological control that inspired both of the authors and many others to pursue studies in this area. Van den Bosch, Hagen, and Huffaker were leaders at the Division of Biological Control at the University of California at Berkeley. The contributions of these three and others at the Division of Biological Control to the science of using natural enemies in integrated control systems foreshadowed today's interest in biologically based pest management systems and provided a solid foundation for biological control for the future.

Mary Louise Flint
Steve H. Dreistadt

Kenneth S. Hagen

Robert van den Bosch

Carl B. Huffaker

Acknowledgments

THIS BOOK WAS PRODUCED under the auspices of the University of California (UC) Statewide IPM Project, Frank G. Zalom, Director, and prepared by IPM Education and Publications of the Statewide IPM Project at the University of California, Davis, Mary Louise Flint, Director.

■

TECHNICAL EDITOR

Elizabeth E. Grafton-Cardwell
Department of Entomology
UC Riverside

■

TECHNICAL ADVISORS

Robert L. Bugg
UC Sustainable Agriculture Research and Education Program
UC Davis

Donald L. Dahlsten
Center for Biological Control
UC Berkeley

Lester E. Ehler
Department of Entomology
UC Davis

Richard D. Goeden
Department of Entomology
UC Riverside

Kenneth S. Hagen
Center for Biological Control
UC Berkeley

Phil A. Phillips
Cooperative Extension and
UC Statewide IPM Project
Ventura County and South Coast Region

Carolyn Pickel
Cooperative Extension and
UC Statewide IPM Project
Sutter-Yuba Counties and
Sacramento Valley

Charles H. Pickett
Biological Control Program
California Department of Food
and Agriculture

Cheryl A. Wilen
Cooperative Extension and
UC Statewide IPM Project
San Diego County and Southern Region

TECHNICAL REVIEWERS AND CONTRIBUTORS

Dave Adams
Joe C. Ball
William W. Barnett
J. Ole Becker
Walt Bentley
William E. Chaney
Donald A. Cooksey
Michael J. Costello
Kent M. Daane
Clyde L. Elmore
Richard Eng
Donald L. Flaherty
Richard Garcia
D. Ken Giles
Peter B. Goodell
Marcella Grebus
Ann E. Hajek
Joseph G. Hancock
Rachid Hanna
Michael P. Hoffmann
Marjorie A. Hoy
John N. Kabashima
Harry K. Kaya
Ann I. King
W. Thomas Lanini
Rachael Long
Robert F. Luck
James D. MacDonald
Mark A. Mayse
Michael V. McKenry
Nicholas J. Mills
Joseph G. Morse
Julie P. Newman
Earl R. Oatman

Patrick J. O'Connor-Marer
Timothy D. Paine
Michael P. Parrella
Albert O. Paulus
Michael J. Pitcairn
Robert D. Raabe
Richard A. Redak
Richard E. Rice
Karen L. Robb
Mike Rose
Robin L. Rosetta
James J. Stapleton
Steve Stauffer
Pavel Svihra
Steven A. Tjosvold
Serguei V. Triapitsyn
Charles E. Turner
Lucia Varela
Stephen C. Welter
Frank G. Zalom
Robert L. Zuparko

SPECIAL THANKS

Max Badgley
Jack S. Bacheler
Peg Brush
Janet Caprile
Johnnie L. Eaton
Joanne Engle
Maxine Hagen
Richard W. Hall
David Headrick
Mark Hoddle
Saralyn K. Huffaker
Jason A. Joseph
Christine Joshel

Shawn King
Steven Lock
Suzanne Paisley
Rosalind R. Rickard
Jay A. Rosenheim
Heinrich Schweizer
Ann Senuta
Larry L. Strand
Will Suckow
Nawal Theodosy
Peggy van den Bosch
Baldo Villegas
Doug E. Walker
Sandra Willard
Cecilia Young

PRODUCTION

Design and Production Coordination:
Seventeenth Street Studios
Editing: Stephen W. Barnett

ILLUSTRATIONS

Copyrighted material is reprinted with permission as cited in figure and table illustrations. We acknowledge the courtesy of those authors and publishers: American Museum of Natural History, Howell V. Daly, Dover Publications, Entomological Society of Canada, Entomological Society of Washington, Henri Goulet, John T. Huber, Minister of Public Works and Government Services Canada, Helen H. Peterson, Queensland Museum, University of California Press, and University of Hawaii Press.

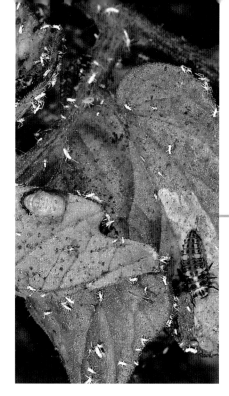

Contents

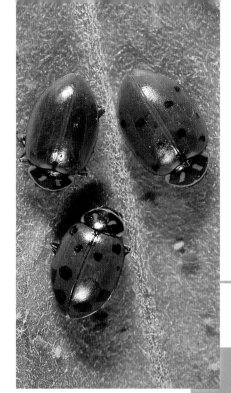

CHAPTER ONE

NATURAL ENEMIES ARE YOUR ALLIES

Consult the Biological Pest Control Quick Guide on page 13 or the index at the back of this book if you are uncertain where to find the information that you want.

Magnifying glass signifies that subject is less than 1 mm.

THIS BOOK IS AN illustrated guide to the identification and biology of beneficial organisms that control pests. Growers, pest control advisers, landscape professionals, home gardeners, pest management teachers and students, and anyone fascinated by natural enemies and their prey will find this book useful. It is meant to be a practical guide focusing on common natural enemies that growers and gardeners are likely to be able to find, identify, and use on almost any crop or in the garden and landscape.

Biological control has been most successful in controlling arthropods (invertebrates such as insects and mites that have jointed appendages and a hard outer skin), and natural enemies of arthropods are emphasized here. Biological control of nematodes, plant pathogens, and weeds is summarized. Discussion of theoretical concepts of biological control has been kept to the minimum necessary to understand how natural enemies function in the field. Although some methods under development are discussed briefly, the reader is referred to scholarly reviews for more information (such as DeBach and Rosen 1991; Lumsden and Vaughn 1993; Mackauer, Ehler, and Roland 1990; Nechols et al. 1995; van den Bosch, Messenger, and Gutierrez 1982; van Driesche and Bellows 1996).

Biological control alone is effective against certain pests, but several methods usually must be combined in an integrated pest management (IPM) program for optimum pest control. Examples of chemical, cultural, and physical management practices in combination with biological control are presented throughout this book.

This publication emphasizes the traditional field of biological control: predators, parasites, and pathogens (Garcia, Caltagirone, and Gutierrez 1988). Crops, rangelands, gardens, and landscapes are the focus. Biological control is important in many other situations, including animal husbandry, forestry (Van Driesche, Healy, and Reardon 1996), postharvest handling (Bull, Stack, and Smilanick 1997), stored products (Flinn, Hagstrum, and McGaughey 1996; Quarles 1996a), structures (Kelley-Tunis, Reid, and Andis 1995), and wildlands, but these situations are not detailed here. Other biologically based, nonchemical pest control methods are sometimes included within broader definitions for biological control (Barbosa and Braxton 1993). These approaches include using pest-resistant plants, improved genetics, pheromones and other behavior-modifying chemicals, and cultural practices; these methods are briefly covered where they can be used to enhance natural enemy activities.

WHAT ARE NATURAL ENEMIES?

Natural enemies are organisms that kill, decrease the reproductive potential, or otherwise reduce the numbers of another organism. In pest management, natural enemies are of interest because they can limit pest damage. The natural enemies discussed in this book reduce pest populations primarily through parasitism or predation (and herbivory in the case of weed control). Other beneficial organisms that compete with pests or secrete substances that inhibit pest activities (antibiosis) can be included in the broad definition of "natural enemy."

■ **Predation.** A predator is an organism that attacks, kills, and feeds on several or many other individuals (its prey) in its lifetime. Some predators are quite specialized and feed on only one or a few closely related species, but most predators are more generalized and feed on a variety of similar organisms. Predators of economic importance in the control of insect pests include beetles (in the insect order Coleoptera), bugs (Hemiptera), flies (Diptera), lacewings (Neuroptera), and wasps (Hymenoptera). Spiders (class Araneae, order Arachnida) can also be important predators of insects. Mites in the family Phytoseiidae are very important in the control of pest mites and certain insects. Mites, collembola, flatworms, protozoa, and other nematodes are among the most important predators of pest nematodes. Giant amoebae and other soilborne animals are pathogens of fungi and bacteria. Birds, bats, fish, and other vertebrates can be important predators of insects and other pests in

Pheromone (sex attractant) is being released by this plastic rope as part of a research project to disrupt adult codling moth mating. This biologically based method is compatible with a selective granulosis virus that kills larvae and with releases of *Trichogramma* parasites to kill moth eggs.

LENGTH ⊢

The tiny size of many natural enemies is illustrated by this shiny black lady beetle on an almond. Many parasites are much smaller than this predator, which is called the spider mite destroyer (*Stethorus picipes*). A microscope is needed to see bacteria and fungi that can control certain pests.

certain situations. Larger carnivores, including raptors, cats, and coyotes, are predators of rodents, pest birds, and other vertebrates.

■ **Parasitism.** A parasite is an organism that lives and feeds in or on a larger host. Unlike predators, parasitic organisms have a prolonged and specialized relationship with their hosts, usually parasitizing only one host individual in their lifetimes. Common parasites of many pests include disease-producing bacteria, fungi, protozoa, viruses, and some nematodes. Parasites may attack invertebrates (such as insects and nematodes), weeds, pathogens, or vertebrates. Parasites may weaken hosts but usually do not kill them; in some cases, however, they have little negative impact. Those that kill or significantly weaken their hosts are natural enemies of importance in biological control.

A parasitic microorganism that causes disease, impairing the normal activities of host tissues or cells, is called a *pathogen*. Disease-causing microorganisms include bacteria, fungi, protozoa, and viruses. Parasitic nematodes are also considered to be pathogens when they cause disease. Pathogens are important in the biological control of many pests, including insects, nematodes, mites, weeds, and other pathogens.

An insect that parasitizes and kills other invertebrates is called a *parasitoid;* parasitoids are parasitic only in their immature stages and kill their host just as they reach maturity. Unlike many other parasites, adult parasitoids are free-living. While other parasites are usually much smaller than their host, adults of many species of parasitoids are about the same size as their host. Most parasitoids are in the insect orders Hymenoptera (wasps) and Diptera (flies); they are very important in biological control of insects and certain other invertebrates. In this book we use the term parasite to include parasitoids.

■ **Herbivory.** Herbivores are animals that feed on plants. Herbivores are important natural enemies of weeds, especially when they specialize and feed on only one or several closely related weedy species. Many of the most effective natural enemies of weeds limit weed reproduction, such as by feeding on flowers or seeds.

■ **Competition.** Competition occurs when two or more organisms strive to obtain the same limited resources (such as food, water, shelter, or light). Competitors with superior ability to obtain resources can be useful in biological control if they have little negative impact on the crop or other managed resource. Competition is important in limiting some weeds and in the success

This adult *Hyposoter exiguae* wasp is laying its egg in a beet armyworm. The parasite egg will hatch into a wasp larva, which will feed inside and kill the caterpillar.

LENGTH

of beneficial microorganisms applied to manage certain pathogens; however, the practical application of competition in biological control is still limited.

■ **Antibiosis.** Some organisms secrete substances that inhibit vital activities of other organisms. For example, many bacteria and molds secrete antimicrobial substances (for example, antibiotics) that inhibit growth of other microorganisms. These substances can be useful in controlling plant pathogens. Allelopathy, which involves the release by one plant of substances that are toxic to another plant species, may have some future application in the biological control of weeds. Although research on antibiosis is rapidly expanding our knowledge of these potential biological control agents, there are currently only a few practical applications for using natural enemies that reduce pests through antibiosis.

■

WHAT IS BIOLOGICAL CONTROL?

Biological control is any activity of one species that reduces the adverse effects of other species. Living natural enemies are the agents of biological control. The impact of a biological control agent results from interactions among populations of the pest and nat-

ural enemy species. A *population* is the local group of individuals of the same species that interbreed, reproduce, feed, and die. Population size (the number of individuals) changes due to interactions among populations and factors such as weather, food, and migration. High populations of many pests will damage crops, whereas low populations may be tolerable. If predators eat pests at a faster rate than pest individuals are born or migrate in, predation reduces the pest population size. If predators reduce pest numbers such that crop loss is reduced and people benefit, this biological control is an important component in the management of that pest (fig. 1-1).

■

USING BIOLOGICAL CONTROL

Virtually every pest has natural enemies that reduce its populations under certain circumstances. Populations of many potential pests are kept below damaging levels most of the time by biological control agents that are part of a local community of organisms. These naturally occurring biological control agents often go unnoticed by pest managers until their activities are disrupted (often by a pesticide or unrecognized factors such as weather), and a new organism becomes a pest (fig. 1-2).

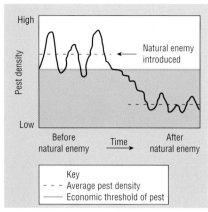

■ FIGURE 1-1. Average pest population size declined after a natural enemy became established. The number of pests is now below the economic threshold or damaging density, resulting in biological control. This illustration applies to classical biological control and inoculative augmentation discussed later.

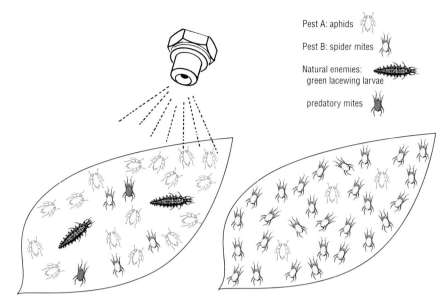

A pesticide applied to control pest A also kills natural enemies that are controlling pest B.

Released from the control exerted by natural enemies, pest B builds up to economically damaging levels.

■ FIGURE 1-2. The destruction of natural enemies often results in secondary outbreaks of insects and mites. For example, spider mites are often present on plants at low densities but become excessively abundant and cause damage when pesticides applied against other species kill the natural enemies of these formerly innocuous species. Here a pesticide applied to kill aphids (Pest A), not only killed aphids, but also killed predaceous green lacewing larvae and predatory mites, leading to a secondary pest outbreak of spider mites (Pest B).

Learning to recognize and identify the naturally occurring natural enemies in a managed system is essential for taking full advantage of biological control. Where naturally occurring biological control is not adequate for maintaining a pest below levels that cause losses, biological control can sometimes be increased by using classical biological control, augmentation, or conservation and enhancement (fig. 1-3).

■ **Conservation and Enhancement.** The majority of the biological control agents at work in our agricultural and urban environments are naturally occurring ones that provide excellent regulation of many potential pests with little or no assistance from people. These agents may include native species or exotic species imported and successfully established as part of a classical biological control program. The existence of naturally occurring biological control agents is one reason that many plant-feeding insects do not ordinarily become economic pests. The importance of such agents often becomes apparent when pesticides applied to control a key pest destroy natural enemies of another species, which then causes an economically destructive secondary pest outbreak (fig. 1-2). Although most

frequently documented for insects and mites (Croft 1990), secondary pest outbreaks occur with other types of pests as well. For instance, pesticides that affect microbial communities in soil or on plant surfaces often increase plant disease problems, suggesting the important role of native microflora in naturally occurring biological control of plant pathogens (Schroth, Hancock, and Weinhold 1992). Shifts in weed species due to repeated herbicide use are a similar example.

Conservation and enhancement offer the greatest opportunities for growers and gardeners to increase their reliance on natural enemies. Eliminating or reducing the use of pesticides toxic to natural enemies is an important way to improve biological control. However, there are many other ways that the local environment can be modified to conserve beneficial species or enhance their activities. Frequently, monoculture systems do not favor natural enemy species. A lack of plant diversity may

increase a crop's susceptibility to pest attack yet reduce its attractiveness to natural enemies. For good survival, natural enemies often require shelter, alternative food sources, overwintering sites, and other conditions not provided in the managed system. Manipulating the environment to improve biological control requires careful consideration of complex interactions. Despite relatively limited research on the effectiveness of habitat manipulation, many growers and landscapers are experimenting with cover crops, alternate hosts, insectary plants, and soil supplements in efforts to conserve native natural enemies.

■ **Augmentation.** Augmentation supplements the numbers of naturally occurring natural enemies with releases of laboratory-reared or field-collected natural enemies. Two approaches to augmentation are inoculative release and inundative release.

Inoculative programs are used to build up populations of a natural enemy

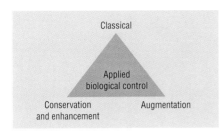

■ FIGURE 1-3. Classical biological control, augmentation, and conservation and enhancement are the three tactics for using natural enemies.

earlier in the season than usual or to establish a natural enemy in orchards or fields where it is not present, often because populations were destroyed by pesticide applications, unfavorable weather, or cultural activities. Once releases are made, the natural enemy is expected to reproduce and increase its populations on its own, at least for that growing season. Examples of inoculative augmentation include release of predatory mites in almonds (Hoy 1984) and strawberries (Strand 1994, Trumble and Morse 1993), numerous programs for citrus insect pests in the Fillmore Protection District (Graebner, Moreno, and Baritelle 1984), and releasing mosquito fish (*Gambusia affinis*) into rice fields and other temporary bodies of water to control mosquitoes (Bay et al. 1976; Kramer, Garcia, and Colwell 1988).

In inundative or periodic release programs, the introduced natural enemies are not expected to reproduce and increase in numbers; biological control is achieved through the activities of the released individuals. Additional releases may be required throughout the season if pest populations approach damaging levels. A wide variety of predators and parasitoids have been used in inundative programs; some of the most well known include the use of *Trichogramma* wasps for control of caterpillars in vegetable crops (such as Oatman et al. 1983), *Aphytis melinus* releases for control of California red scale (Haney et al. 1992, 1993; Moreno and Luck 1992), and release of leafminer and whitefly parasites in greenhouses (Parrella et al. 1991, van Lenteren and Woets 1988). Inundative releases of entomopatho-

genic nematodes are used against a number of insect species including black vine weevil and carpenterworm and other borers (Gaugler and Kaya 1990). These nematodes are called entomopathogenic because they carry a bacterium that is pathogenic to insects, and the bacterium in combination with the nematode kills the insect host.

Inundative release is currently the primary way microorganisms are used in biological control. Microorganisms must be registered as pesticides prior to release in the environment. Microbial pesticides are used against insects, weeds, and plant pathogens and have potential for use against nematodes.

■ **Importation or Classical Biological Control.** Classical biological control involves the deliberate introduction and establishment of exotic natural enemies into areas where they did not previously occur. Such programs are employed largely against pests of foreign origin that have accidentally become established and are able to develop to high numbers in the absence of the natural enemies that keep them under control in their native areas. These programs have been successfully used primarily against insect pests and weeds. Classical biological control is conducted by specially trained university and government agency personnel in quarantine facilities certified to handle exotic organisms (Van Driesche and Bellows 1993).

Several steps are required to implement a successful importation program. First, the pest's native area must be identified, which requires accurate identification of the pest organism itself. Then a search of the native area is carried out, and promising natural enemies are shipped back for testing. Shipments must be held and studied in an approved quarantine facility to exclude potential contaminants and to confirm that the natural enemy will have minimal negative impact in the new country of release (Coulson, Soper, and Williams 1991). After appropriate quarantine processing and biological testing, the natural enemies are usually reared

to increase their numbers and are then released into the new area. The goal is to permanently establish the natural enemy species so that it can provide a regulating influence on the pest population (fig. 1-1). Several repeated releases or reimportations of natural enemies may be necessary. Often, imported natural enemies do not thrive in the new area, and other biotypes or species that are better adapted to the local environment must be introduced. Such persistence has paid off: at least 63 invertebrate pests in the United States and reportedly more than 400 different insect species worldwide (Greathead and Greathead 1992, Laing and Hamai 1976, Luck 1981) have been partially, substantially, or completely controlled by classical biological control. Complete control means that the pest normally requires no supplementary management if the activities of the natural enemy are not disrupted. Numerous weeds have also been managed with classical biological control, with 14 weeds partially or completely controlled in California alone (Turner et al. 1992). Many more insect (Flint et al. 1992) and weed pests are currently considered to be candidates for biological control importation programs.

Classical biological control is carried out only by government agencies or university experiment stations and is not an option for individual growers or pest managers. However, it is important for pest managers to recognize imported natural enemies and to conserve them when possible. Because classical biological control can provide long-term benefits over a large area and is conducted by agencies and institutions funded through taxes, support from growers, pest control professionals, and the public is critical to the continued success of classical biological control.

■

IS BIOLOGICAL CONTROL "SAFE"?

One of the great benefits of biological control is its relative safety for human health and the environment. However,

Birds such as house finches eat insects and can be important natural biological control agents of weeds, consuming many seeds. Unfortunately, they also flock in fields and feed on some crops. Omnivorous feeders, such as finches and most other birds, are not used for applied biological control.

FIVE KEY COMPONENTS OF IPM PROGRAMS

- preventing pest problems;

- identifying pests and learning their ecology;

- regular monitoring of plants, potential pests, and natural enemies and the conditions and practices that affect them;

- use of control action thresholds and guidelines; and

- integrating the use of biological, chemical, cultural, and physical control methods.

introduction of certain exotic organisms has sometimes adversely impacted native or beneficial species. Most negative impacts from exotic species have been caused when undesirable organisms contaminate imported goods or when non-scientists, such as travelers, bring in pest-infested fruit or ornamental plants that become weeds. These illegal importations are not part of biological control programs.

Deliberate introduction and use of certain natural enemies, such as those with broad host ranges, may adversely affect native or desirable species (Horn 1996, Howarth 1991, Johnson et al. 1995, Johnson and Stiling 1996). However, instances of this are relatively uncommon and scientists disagree about whether they are significant (Gould, Kennedy, and Kopanic 1996).

Negative impacts have occurred from poorly conceived, "quasi-biological control" importations of predaceous vertebrates, such as mongooses and certain fish.

To avoid these problems, biological control researchers follow strict government quarantine regulations (Goeden 1992) and work with relatively host-specific natural enemies. Introductions of more general feeders, such as certain lady beetles and *Trichogramma* spp. wasps, are more controversial because these can attack beneficial or innocuous insects as well as the target pests. Biological control provides great benefits. The environmental and public health risks from careful biological control projects are relatively low (Anonymous 1995a; Cook et al. 1996; Greathead 1995; Hokkanen and Lynch 1995).

USING NATURAL ENEMIES WITHIN AN INTEGRATED PEST MANAGEMENT PROGRAM

Although natural enemies sometimes provide sufficient control to completely manage a pest, it is common for natural enemies to reduce a pest population substantially but not enough to prevent economic loss all the time. For reliable management and to provide optimal pest control, many natural enemies must be employed along with other management methods, such as host-plant resistance, cultural practices, and compatible chemical controls.

The integrated control concept was developed by entomologists to combine the use of natural enemies with insecticides for more reliable control (Stern et al. 1959). This integrated control concept was expanded into integrated pest management (IPM) to include the management of all types of pests using a variety of compatible control tools. Reliance on biological control is a basic tenet of IPM: "An ecologically based pest control strategy that relies heavily on natural mortality factors . . . and seeks out control tactics that disrupt these factors as little as possible" (Flint and van den Bosch 1981, 6).

Integrated pest management is an ecosystem-based pest management strategy that seeks to prevent and suppress pest problems using a combination of compatible techniques with a minimum of adverse impacts on human health, the environment, and nontarget organisms.

Although IPM is often thought of as a way to use pesticides more successfully, employing all the components of IPM can vastly improve the effectiveness of natural enemies in our managed systems.

■ **Prevention.** Because biological control agents often suppress pests before they become a problem, natural enemies are an important prevention component of many IPM programs. Natural enemies whose activities and population growth are well synchronized with those of the pest—so that harmful pest populations rarely occur—are often regarded as ideal biological control agents because they prevent pest outbreaks. Other natural enemies often do not become abundant until pests rise to damaging levels, too late to prevent economic damage. Augmentative release of natural enemies is usually most effective when enough pests are present to support the introduced natural enemies but pest populations remain below the level that would require a pesticide treatment.

■ **Identifying Pests and Ecological Factors.** Proper pest identification is essential for maximum use of biological control. Many of the most effective natural enemies control only one species or several closely related pest species. Improper pest identification can lead to incorrect conclusions about the effectiveness or potential of a natural enemy. A good understanding of how pest biology or activities are affected by weather and other environmental factors can also help in evaluating the need for management and the impact of biological control agents.

■ **Monitoring.** Field monitoring provides information on the crop, pests, and biological control agents. Monitoring tools and techniques vary according to the managed system and target pest. For insect pests, actual counts of insects on leaves or in traps may be appropriate, whereas soil samples are a common

Use a 10 power hand lens to detect and identify mites and small insects. Hold the lens near your eye and move the object being viewed close until it is in focus.

monitoring method for nematodes. Plant pathogens can be monitored with soil or plant assays or by host-plant damage symptoms. To assess weed problems, weed seeds, seedlings, or mature plants may be sampled at various times of the year. Monitoring of weather, moisture, and other environmental conditions and cultural practices are important in helping to assess the potential for many future pest problems.

Monitoring of natural enemies is a key component of an IPM program and is essential for getting a maximum understanding of, and benefit from, biological control. Monitoring may involve sampling for the natural enemy itself or for diseased, parasitized, or killed host pests. Monitoring methods for natural enemies of a few insects and mites have been well established; for other natural enemies, you must devise your own methods. (Examples of monitoring methods for insects and mites are given in table 6-4.) Keep written monitoring records over time. During the season these records can show whether pest or natural enemy populations are rising or decreasing, and records can be compared with other seasons. These monitoring records may help to forecast possible outbreaks of the next generation of a pest and the potential for biological control.

Pest population counts, together with records of natural enemies, control measures, cultural practices, weather conditions, and crop yield or plant development and health, can help you decide whether future control actions will be needed—including augmentative release of biological control agents. Making simple tables and graphs for the data you collect will help you identify population patterns and areas of pest hot spots or natural suppression. Over the years, these records provide valuable historical data for long-term management.

■ **Control Action Thresholds and Guidelines.** Control action thresholds or guidelines tell you when management actions are needed to avoid losses due to pests. Such guidelines are formulated on the IPM concept that not every field or landscape needs to be treated every year for every pest. Guidelines for some pests, especially insects and mites, include numerical thresholds based on specific sampling or monitoring techniques; for these pests, there is an understanding that they may be tolerated at low population levels, but perhaps not when their numbers get high. Guidelines for other pests, including most pathogens and weeds, are based on field history, stage of crop development, presence of symptoms or damage, weather conditions, and other observations. Although often used to determine the need for pesticide treatments, control action guidelines may also tell growers when an augmentative release of natural enemies is required. In an IPM program that truly integrates the use of naturally occurring natural enemies, control action thresholds incorporate not only measures of pest numbers but also measures of biological control agents that may keep pest numbers from reaching intolerable levels (table 1-1).

For example, in processing tomatoes in the Sacramento Valley, the need for treating tomato fruitworm depends on the level of parasitization of the pest eggs by the tiny wasp *Trichogramma pretiosum* and by the number of fruitworm eggs per leaf. Pesticide treatment is not needed if there is at least 1 visibly parasitized (black) egg for every 8 white eggs on a leaf containing 8 or fewer fruitworm eggs (Hoffmann et al. 1991b; Toscano, Zalom, and Trumble 1995). Acceptable parasitization rates change with the density of fruitworm eggs, as shown in figure 1-4. Although there are relatively few research-based control action guidelines incorporating monitoring of natural enemies, many managers develop their own "seat of the pants" guidelines using records of natural enemy and pest occurrences in previous seasons.

■ **Integrating Compatible Management Tools.** Integrated pest management relies on a variety of cultural,

TABLE 1-1. Examples of pest control action guidelines based on natural enemy abundance.

PEST	CROP	NATURAL ENEMY	THRESHOLD	REFERENCE
alfalfa caterpillar	alfalfa	*Cotesia medicaginis*	10 or more unparasitized caterpillars per net sweep	Summers, Hagen, and Stern 1996
alfalfa looper and cabbage looper	cotton	various predators *Copidosoma truncatellum* *Hyposoter exiguae* *Microplitis brassicae* *Trichogramma pretiosum* *Voria ruralis*	avoid treating moderate populations, which are beneficial prey for natural enemies of other pests	Anonymous 1996
tentiform leafminers	apple	*Pnigalio flavipes* *Sympiesis stigmata*	2–5 or more mines per leaf and <10–30% parasitism, depending on the pest generation	Anonymous 1991a, Caprile et al. 1996
California red scale	citrus	*Aphytis* spp. *Comperiella bifasciata*	percentage of fruit with 1 or more live, unparasitized scales	Grafton-Cardwell et al. 1996
citrus red mite	citrus	*Euseius stipulatus* *Euseius tularensis*	if <1 predator mite for every 1–2 pest mites	Grafton-Cardwell et al. 1996
citrus aphids, including spirea aphid and cotton or melon aphid	citrus	many species	important spring food for natural enemies of other pests; no treatment if ≤40% of growth flushes are infested	Grafton-Cardwell et al. 1996
citrus thrips	citrus	*Anystis agilis* dustywings *Euseius tularensis* lacewings minute pirate bugs	if predators >0.5 per leaf, treat navel orange if 10% of fruit are thrips-infested; treat at 5% fruit infestation if fewer predators are present	Grafton-Cardwell et al. 1996, Haney et al. 1992
European fruit lecanium frosted scale	walnut	*Metaphycus californicus*	if during dormant season >5 nymphs/ft (30 cm) of previous year's growth and <90% are parasitized	Hendricks et al. 1996
Fuller rose beetle	citrus	*Fidiobia citri*	if >0.1% of fruit are infested with unparasitized, unhatched eggs	Grafton-Cardwell et al. 1996
grape leafhopper	grape	*Anagrus* spp.	if parasites are active on first-generation eggs, avoid treating raisin or wine grapes unless leafhopper nymphs are well above 20/leaf	Bentley et al. 1996
spider mites	almond	western predatory mite	no treatment, or rates as low as ⅛ to ⅒ of label rate, based on predator/prey ratio	Zalom et al. 1996
spider mites	apple, pear	western predatory mite	no treatment if 1 or more predator per 10 prey	Bethell et al. 1995, Caprile et al. 1996
spotted alfalfa aphid	alfalfa	*Hippodamia convergens* and other lady beetles	if <1 lady beetle per 5–50 aphids, varying with predator and crop stage	Summers, Hagen, and Stern 1996
tomato fruitworm	tomato	*Trichogramma* egg parasites	ratio of parasitized to unparasitized eggs, as in fig. 1-4	Toscano, Zalom, and Trumble 1995
whiteflies	citrus	parasitic wasps	do not treat; insecticides cause resurgence; conserve parasites	Grafton-Cardwell et al. 1996
western grapeleaf skeletonizer	grape	granulosis virus	if virus is not present, treat after bloom when larvae are found	Bentley et al. 1996
woolly apple aphid	apple	*Aphelinus mali*	treat only if parasites have been disrupted, typically by insecticides applied for other pests	Caprile et al. 1996

PEST	CROP	NATURAL ENEMY	THRESHOLD	REFERENCE
western flower thrips	cotton, cucurbits, dry beans, figs, strawberry	avoid treating western flower thrips; it is a beneficial spider mite predator and alternate prey for generalist predators		Anonymous 1996; Coviello, Bentley, and Barnett 1996; Godfrey 1995; Godfrey et al. 1996

Note: These examples are from University of California *UC IPM Pest Management Guidelines.* Only a summary of control action guidelines is presented here. Consult the references for sampling methods and details before using these thresholds. Consult the index for common and scientific names of the invertebrates mentioned here.

mechanical, physical, and chemical control methods as well as biological control agents. Selection of management tools must include consideration of how they might impact other pest problems or biological control agents. IPM programs often combine two or more compatible methods that when used alone might not provide adequate control but when used together provide excellent management. Preferred pest management techniques include encouraging naturally occurring biological control; using alternative plant species or varieties that resist pests; adoption of cultivating, pruning, fertilizing, or irrigation practices that reduce pest problems; or changing the habitat to make it incompatible with pest development. When pesticides are needed, the least toxic but effective material should be chosen with special consideration of its potential impact on natural enemies and other pests in the managed ecosystem. Examples of how chemical, cultural, mechanical, and physical control methods can be used in conjunction with biological control agents in integrated pest management programs are described below.

■ **Chemical and Biological Control Integration.** Pesticides are chemicals that control, prevent, repel, or mitigate pests or the problems they cause. You can quickly obtain temporary control of certain pests if you choose the correct pesticide and apply it at the right time in an appropriate manner. Incorrect pesticide use, such as applying the wrong material, using the wrong rate,

or applying by an improper method, can do more harm than good, such as damaging natural enemies. Secondary pest outbreaks (fig. 1-2), target pest resurgence (fig. 1-5), accelerated pesticide resistance (fig. 1-6), and increased pesticide use, hazard, and environmental contamination are caused by pesticide uses incompatible with natural enemies. Always understand the relative toxicity, mode of action, persistence, and safe and legal use of pesticides you consider using (Marer, Flint, and Stimmann 1988), especially their human and environmental safety, selectivity, and potential impact on biological control.

There are many methods for integrating chemical and biological control (Dreistadt, Parrella, and Flint 1992;

Wright and Verkerk 1995). A selective mode of action allows biological insecticides or microbials, in particular several varieties of *Bacillus thuringiensis* (Bt), to be applied against target pests (certain Lepidoptera, Coleoptera, or Diptera) without direct toxicity to most other organisms (table 1-2). The slow-acting stomach poison cryolite controls caterpillars, katydids, and Fuller rose beetles infesting citrus with little or no adverse effect on the many natural enemies important in controlling various citrus pests. Entomopathogenic nematodes combined with Bt or insecticidal soap can be applied to plants infested with a complex of soil- and foliage-feeding pests, including black vine weevil, cabbage aphid, cabbage looper, cucumber beetles, masked chafer, and

BLACK EGGS	NUMBER OF WHITE EGGS							
	4–8	9	10	11	12	13	14	15
1	-	T	T	T	T	T	T	T
2	-	-	-	T	T	T	T	T
3	-	-	-	-	T	T	T	T
4	-	-	-	-	-	T	T	T
5	-	-	-	-	-	T	T	T
6	-	-	-	-	-	T	T	T
7	-	-	-	-	-	-	T	T
8	-	-	-	-	-	-	T	T
9	-	-	-	-	-	-	-	T
10	-	-	-	-	-	-	-	T

■ FIGURE 1-4. The ratio of parasitized (black) to unparasitized (white) tomato fruitworm eggs used to determine whether treatment is warranted in Sacramento Valley processing tomatoes. The letter T indicates the ratio at which treatment is recommended. See the *Tomato Pest Management Guidelines: Insects and Mites* (Toscano, Zalom, and Trumble 1995) for the recommended sampling method and other details.

Aphid (pest)

Lacewing larva (predator)

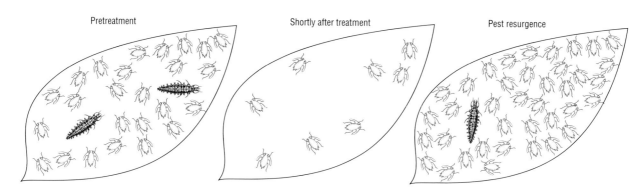

Pretreatment Shortly after treatment Pest resurgence

■ FIGURE 1-5. Target pest resurgence can result when natural enemies are destroyed. In comparison with their effect on pest species, pesticides often kill a higher proportion of the populations of predators and parasites. The immediate effect of spraying is not only a reduction in the number of pests, but an even greater reduction of the natural enemies. The resulting unfavorable ratio of pests to natural enemies permits a rapid increase or resurgence of the pest population.

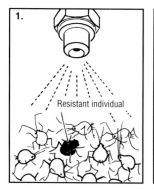

1. Resistant individual

2.

3. Susceptible individual

■ FIGURE 1-6. Pest populations develop resistance to pesticides through genetic selection (Anonymous 1986): 1: Certain individuals in a pest population are less-susceptible to a pesticide spray than other individuals. 2: These less-susceptible pests are more likely to survive an application and to produce less-susceptible progeny. 3: After repeated applications, the pest population consists primarily of resistant or less-susceptible individuals, and applying the same material, or other chemicals with the same mode of action, is no longer effective.

white grubs (Kaya et al. 1995, Koppenhöfer and Kaya 1997). In comparison with more persistent materials, horticultural oils (Davidson et al. 1991) or insecticidal soaps have little or no residual toxicity to natural enemies migrating in or emerging after an application. Systemic insecticides can be applied to avoid exposing natural enemies directly; the indirect effects of systemics on natural enemies that feed on plant parts (such as nectar and pollen) or treated pests is largely unknown.

Selective pesticide placement or timing of applications can take advantage of pest behavior and biology to minimize natural enemy exposure (Croft 1990) and can even enhance biological control (table 1-3). For example, psyllid parasites pupating within hosts (Mensah and Madden 1993) and some green peach aphid parasites within aphid mummies (Hsieh and Allen 1986) are resistant to certain insecticides. Dormant-season treatments may minimize exposure of natural enemies that are in an overwintering state, especially if natural enemies overwinter away from their host plants. However, many parasites overwinter on host plants or as immatures inside their prey, and there is little data showing that dormant season sprays minimize adverse effects on natural enemies. Some broad-spectrum materials can be selectively applied by incorporation into baits. Enclosed baits placed around the plant base or insecticides applied to soil around plants or to the trunk can control pest-tending ants, thereby improving natural enemy effectiveness.

■ **Cultural and Biological Control Integration.** Cultural controls are modifications of normal plant care activities that reduce or avoid pest problems. Proper planting and good care, such as appropriate irrigation, can prevent many pests from adversely affecting plants. Cultural care also affects the relationship between pests and their natural enemies. Excess fertilization can overstimulate succulent plant growth, prompting populations of aphids, mites, psyllids, and certain other pests to develop rapidly and at least temporarily escape regulation by their natural enemies. Planting time; cultivation time, method, and frequency; harvesting time and method; crop rotation; and sanitation are cultural methods that can influence the effectiveness of biological control and may be included within the definitions of augmentation or conservation methods of biological control. Various living mulches (Lanini, Shribbs, and Elmore 1988) or cover crops are used to improve pest management.

There are many examples of cultural practices that enhance or disrupt biological control. Alternate-row pruning (pruning half the rows one month and trimming the rows in between a month or more later) is recommended in citrus because many whiteflies prefer to lay eggs on the succulent new growth that develops after pruning (Anonymous 1991b). Because parasites are removed from the orchard when the older stage whiteflies in which they are developing on older terminals are trimmed, alternate row pruning provides refuges for whitefly parasitoids, allowing them to emerge from older growth on untrimmed trees and attack whiteflies infesting the new growth of nearby, recently trimmed trees.

Pruning of eugenia into shapes (topiary) may reduce density of eugenia psyllids that lay eggs on new growth. However, depending on timing and the handling of pruned tips, pruning can disrupt biological control if parasites developing within psyllid nymphs are removed when older foliage is pruned-off. Pruning the interior of citrus trees increases predaceous mite populations in the exterior canopy, thereby reducing fruit scarring by citrus thrips (Grafton-Cardwell and Ouyang 1995a, 1995b). Strip harvesting or border cropping conserves natural enemies in alfalfa. For more information on habitat manipulation, see the discussion of conservation and enhancement in chapter 6.

■ **Mechanical and Biological Control Integration.** Mechanical controls use labor, materials not usually considered to be pesticides, and machinery to reduce pest populations directly. Barriers to exclude ants are an example of mechanical control; keeping pest-tending ants off plants can increase natural enemy populations and improve biological control.

Screens allow certain pests infesting greenhouse crops to be controlled

TABLE 1-2. Toxicity of common types of insecticides to natural enemies.

INSECTICIDE	TOXICITY	
	Contact	*Residual*
biologicals*	no	no
oil, soap	yes	no
IGRs§	yes/no	yes/no
botanicals‡	yes	no
pyrethroids	yes	yes
carbamates and organophosphates	yes	yes

Note: Contact toxicity is immediate killing resulting from spraying the beneficial or its habitat. Residual toxicity is from residues that persist and kill natural enemies that migrate in and contact previously treated areas. See table 6-2 and figure 6-1 for examples of each type of pesticide and more details on insecticide toxicity to specific groups of natural enemies.

* Biologicals refer to host-specific materials; the most common of these materials are also called microbials and include subspecies of *Bacillus thuringiensis* (Bt). Some pesticides, like Bt and the inorganic cryolite, are very specific because they act only as stomach poisons and are toxic only to certain types of insects that feed on treated plants. Pheromones used for mating disruption are another group of very specific pesticides.

§ IGRs are insect growth regulators, mostly newer chemicals with very insect-specific modes of action. Yes/no indicates that toxicity varies depending on the specific material and the species and life stage of the natural enemy.

‡ Botanicals are pesticides extracted from plants. The botanical sabadilla can have low toxicity to certain beneficials, in part because it is toxic mostly through ingestion. Neem can have low toxicity to some natural enemies, such as mature adults that are not susceptible to neem's growth-regulating effects.

through introductions of natural enemies. Without special screening that excludes insects but allows adequate ventilation, crops would become so heavily infested from pests migrating in that it would not be economical or feasible to release sufficient numbers of natural enemies at the correct time to provide augmentative biological control. Some barriers exclude pests but not natural enemies, allowing beneficials to colonize the crop (see the discussion of augmentation in chapter 6).

■ **Physical and Biological Control Integration.** Physical controls are environmental manipulations that indirectly control pests or prevent damage by altering temperature, light, and humidity. All organisms are adapted to a certain range of environmental conditions and these tolerances vary among organisms. For example, heat applied through solarization or steam pasteurization can be used in combination

Insecticide is being applied as a band encircling the trunk to kill elm leaf beetle larvae that crawl down from the canopy to pupate around the tree base. This selective application method conserves natural enemies in the tree canopy that would be killed if this broad-spectrum, persistent insecticide were sprayed onto the foliage.

TABLE 1-3. Examples of selective placement or spot applications of pesticide that conserve natural enemies.

PESTS	NATURAL ENEMIES	PESTICIDE APPLICATION METHOD	REFERENCE
brown garden snail	*Ruminia decollata*	bait only in a narrow strip in the middle between tree rows	Anonymous 1991b
citrus thrips	*Euseius tularensis* and others	spray only the outer tree canopy	Anonymous 1991b
cutworms infesting field crops	many parasites and predators	apply bait in plant rows	
elm leaf beetle	*Frynniopsis antennata*	bark banding	Dreistadt and Dahlsten 1990a
Homoptera-tending ants	many parasites and predators	enclosed bait, soil or basal trunk spray	Baker, Key, and Gaston 1985; Knight and Rust 1991; Phillips, Bekey, and Goodall 1987
spider mites on apples	predaceous mites and general predators	close bottom nozzles and direct apple-thinning sprays away from tree centers	Anonymous 1991a
spider mites on cut roses	*Phytoseiulus persimilis*	spray only upper canopy marketed parts	Zhang and Sanderson 1995
various insects and pest mites in orchards	spider mite destroyer, predatory mites, and other natural enemies	spray only every other row in an orchard, so only half of trees are treated on each date	Hull and Beers 1985

Note: Consult the index for common and scientific names of the invertebrates mentioned here.

with beneficial microorganisms or amendments to improve control of some soilborne pathogens (Elmore et al. 1997, Gamliel and Stapleton 1993). Invertebrate natural enemies often require warmer temperatures to develop than do pests. This ensures that under natural conditions, pests develop first and sufficient numbers are available to serve as hosts once the natural enemies develop. These different temperature adaptations can result in a lag time during which pests become temporarily abundant before natural enemies appear and reproduce sufficiently to provide control.

It is important for pest managers to be aware of how the environmental conditions at their sites can influence the specific biological controls in their crops. For example, thinning tree canopies in hot areas of California increases scale insect mortality due to heat exposure (Daane and Caltagirone 1989), but the resulting hotter, drier conditions may also adversely affect parasites. Some commercially available predaceous mite species are better adapted than other mite species to particular crops and certain temperatures and humidities (see table 8-5). Predaceous mites, bugs, parasites, and other types of natural enemies that can be effective for pest control when released in greenhouses during the summer are often ineffective during winter; often because many beneficials stop reproducing under short day length or cooler temperatures even though pests are still present.

If releases are planned, choose the species best adapted for the anticipated conditions at that site. If possible, modify conditions to enhance natural enemy survival and reproduction. Temperature and light can readily be modified in most greenhouses. Removing or adding structures, windbreaks, nearby plants, ground covers, and certain mulches change reflected light and radiated heat outdoors. Irrigation timing and method (such as drip versus sprinkler) can be used to alter humidity and temperature.

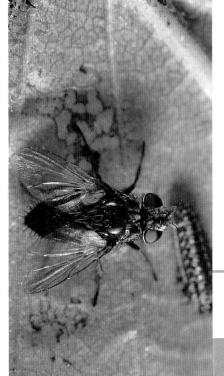

CHAPTER TWO

BIOLOGICAL PEST CONTROL QUICK GUIDE

THIS QUICK GUIDE WILL help you identify the natural enemies most likely to help control pests in your farm, garden, or landscape. Pest arthropods are listed alphabetically according to the common name of major groups, such as aphids, beetles, caterpillars, and mites. Nematodes, plant pathogens, and weeds are each a single group. Below the pest group name (such as aphids or beetles) is a list of "general" predators, pathogens, and other natural enemies that attack many species within that group. Some common pest species within that group are also listed along with the names of host-specific parasites or other natural enemies that attack that pest species but not most other pests. Page numbers refer you to places in this book where those or similar natural enemies are discussed and pictured.

If your pest is listed by species name, *don't forget* to also look at the natural enemies attacking many species in that pest group—many of these can be very important in biological control! If your pest is not listed by species name, look at the pest group name for more general predators and parasites likely to pro-

vide some control. There are probably host-specific parasites or other natural enemies that attack your pest but are not listed here.

This Quick Guide lists natural enemies identified as important in the management of the pests covered in *University of California Pest Management Guidelines*, UC IPM manuals, *Pests of the Garden and Small Farm* (Flint 1990), *Pests of Landscape Trees and Shrubs* (Dreistadt 1994), and natural enemies used specifically as examples in this book. Additional natural enemies were suggested by reviewers of draft manuscripts of this book.

Although many pests and natural enemies that occur in farms and gardens are not included, this Quick Guide provides a basic introduction to important natural enemies that you can recognize. Only a few natural enemies are listed for nematodes, plant pathogens, and weeds because relatively little is known about the practical use of biological control agents against most of these pests. For more information, refer to the literature cited in the discussions of specific natural enemy groups and in the suggested reading section at the end of this book.

Recognize your allies in combating these pests. Find photographs
and the biology of natural enemies on pages listed here.

APHIDS

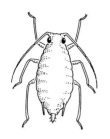

Natural Enemies attacking many aphids	Page
aphid flies (predators)	95
bugs (predators)	91
earwigs (predators)	105
Entomophthora aphidis (fungus)	121
lacewing larvae (predators)	100
lady beetles (predators)	82
soldier beetles (predators)	89
syrphid larvae (predators)	94
wasps (parasites)	66

Specific pest	Natural enemies	Page
bean aphid	*Diaeretiella* spp. (endoparasitic wasps)	66–67
black bean aphid	*Lysiphlebus* spp. (endoparasitic wasps)	67–68
blue alfalfa aphid	*Aphidius ervi* (endoparasitic wasp)	
cabbage aphid	*Diaeretiella rapae* (endoparasitic wasp)	66–67
corn leaf aphid	*Lysiphlebus testaceipes* (endoparasitic wasp)	67–68
cotton or melon aphid	*Lysiphlebus testaceipes* (endoparasitic wasp)	67–68
elm aphid	*Trioxys tenuicaudus* (endoparasitic wasp)	
green peach aphid	*Aphelinus semiflavus* (endoparasitic wasp)	67
	Aphidius matricariae (endoparasitic wasp)	
	Diaeretiella rapae (endoparasitic wasp)	66–67
	Lysiphlebus testaceipes (endoparasitic wasp)	67–68
leaf curl plum aphid	*Aphelinus* sp. (endoparasitic wasp)	67
	Aphidius matricariae (endoparasitic wasp)	
linden aphid	*Trioxys curvicaudus* (endoparasitic wasp)	
mealy plum aphid	*Praon* spp. (endoparasitic wasps)	
pea aphid	*Aphidius ervi* (endoparasitic wasp)	
	Aphidius smithi (endoparasitic wasp)	
rosy apple aphid	*Lysiphlebus testaceipes* (endoparasitic wasp)	67–68
spotted alfalfa aphid	convergent lady beetle (predator)	82
	Trioxys complanatus (endoparasitic wasp)	
walnut aphid	*Trioxys pallidus* (endoparasitic wasp)	67
woolly apple aphid	*Aphelinus mali* (endoparasitic wasp)	8

Note: Consult the index at the end of this book for complete listings and common and scientific
names of the organisms mentioned here.

Endoparasitic species feed inside their host; ectoparasites feed attached to their host's outside
(see chapter 6).

BEETLES

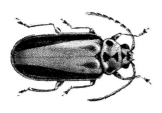

Natural Enemies attacking many beetles	Page
bacteria, fungi, and viruses (pathogens)	117
bugs (predators)	91
flies (parasites)	73
nematodes (entomopathogens)	119
wasps (parasites)	57

Specific pest	Natural enemies	Page
alfalfa weevils	*Bathyplectes* spp. (larval-prepupal endoparasitic wasps)	62–63
	Microctonus aethiopoides (adult endoparasitic wasp)	62–63
	Oomyzus (=Tetrastichus) incertus (larval endoparasitic wasp)	63
	Zoophthora phytonomi (entomogenous fungus)	63
billbugs	*Steinernema carpocapsae* (entomopathogenic nematode)	119
black turfgrass ataenius	*Heterorhabditis bacteriophora* (entomopathogenic nematode)	119
	Steinernema carpocapsae (entomopathogenic nematode)	119
black vine weevil	*Steinernema carpocapsae* (entomopathogenic nematode)	119
Colorado potato beetle	*B. thuringiensis* ssp. *tenebrionis* (pathogen)	118
	Edovum puttleri (egg endoparasitic wasp) twospotted stink bug (predator)	94
Egyptian alfalfa weevil	*Microctonus aethiopoides* (adult endoparasitic wasp)	62–63
elm leaf beetle	*B. thuringiensis* ssp. *tenebrionis* (pathogen)	118
	Erynniopsis antennata (pupal endoparasitic fly)	75–76
	Oomyzus (=Tetrastichus) spp. (egg, pupal endoparasitic wasps)	76
eucalyptus longhorned borer	*Avetianella longoi* (egg parasitic wasp)	64
	Syngaster lepidus (larval parasitic wasp)	
eucalyptus snout beetle	*Anaphes nitens* (egg endoparasitic wasp)	64
Fuller rose beetle	*Fidiobia citri* (egg endoparasitic wasp)	8
Japanese beetle	*Bacillus popilliae* (pathogen)	118
	Heterorhabditis, Steinernema spp. (entomopathogenic nematodes)	119
	Hyperecteina aldrichi (adult endoparasitic fly)	76–77
masked chafers or white grubs	*Heterorhabditis bacteriophora* (entomopathogenic nematode)	119
	Steinernema glaseri (entomopathogenic nematode)	119
Mexican bean beetle	*Podisus maculiventris* (predatory stink bug)	92, 94

BUGS

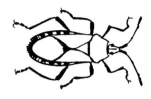

Natural Enemies attacking many bugs	Page
assassin bugs (predators)	93
bigeyed bugs (predators)	93
damsel bugs (predators)	94
flies (parasites)	73
lacewings (predators)	100
wasps (parasites)	57

Specific pest	Natural enemies	Page
bordererd plant bug, squash bug	*Trichopoda pennipes* (adult endoparasitic fly)	73
lygus bug	bigeyed bugs (predators) *Anaphes iole* (egg endoparasitic wasp)	92–93
stink bugs	*Trissolcus basalis, T. euschisti* (egg endoparasitic wasps)	60

CATERPILLARS

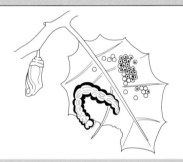

Natural Enemies attacking many caterpillars	Page
assassin bugs (predators)	93
Bacillus thuringiensis (pathogen)	118
bigeyed bugs (predators)	93
birds (predators)	116
damsel bugs (predators)	94
earwigs (predators)	105
flies (parasites)	73
ground beetles (predators)	89
lacewings (predators)	100
minute pirate bugs (predators)	91
nematodes (entomopathogens)	119
Phytocoris spp. plant bugs (predators)	93
spiders (predators)	110
viruses (pathogens)	123
wasps (parasites)	57

Specific pest	Natural enemies	Page
alfalfa looper cabbage looper	*Copidosoma truncatellum* (egg-larval endoparasitic wasp)	62
	Hyposoter exiguae (larval endoparasitic wasp)	61
	Microplitis brassicae (larval endoparasitic wasp)	
	nuclear polyhedrosis virus (pathogen)	123
	Trichogramma pretiosum (egg endoparasitic wasp)	56–60
	Voria ruralis (larval endoparasitic fly)	75
alfalfa caterpillar	*Cotesia medicaginis* (larval endoparasitic wasp)	61
	Trichogramma semifumatum (egg endoparasitic wasp)	56–60
amorbia	*Erynnia tortricis* (larval-pupal endoparasitic fly)	75
	Trichogramma platneri (egg endoparasitic wasp)	56–60
artichoke plume moth	*Steinernema carpocapsae* (entomopathogenic nematode)	120
	Trichogramma spp. (egg endoparasitic wasps)	56–60
beet armyworm yellowstriped armyworm	*Archytas apicifer* (larval-pupal endoparasitic fly)	
	Chelonus texanus (=*C. insularis*) (egg-larval endoparasitic wasp)	
	Cotesia marginiventris (larval endoparasitic wasp)	61
	Hyposoter exiguae (larval endoparasitic wasp)	61
	Lespesia archippivora (larval-pupal endoparasitic fly)	
	Trichogramma spp. (egg endoparasitic wasps)	56–60
	viral disease (pathogen)	123
California orangedog	*Hyposoter* sp. (larval endoparasitic wasp)	60–61
carpenterworm	*Steinernema feltiae* (entomopathogenic nematode)	120

	Specific pest	Natural enemies	Page

	codling moth	*Ascogaster quadridentata*
		  (egg-prepupal endoparasitic wasp)
		Macrocentrus ancylivorus
		  (larval-pupal endoparasitic wasp)
		Trichogramma spp. (egg endoparasitic wasps)
		granulosis virus (pathogen)
			
			56–60
			123

	corn earworm cotton bollworm tobacco budworm tomato fruitworm	*Archytas apicifer* (larval-pupal endoparasitic fly) *Chelonus texanus* (=*C. insularis*) (egg-larval endoparasitic wasps) *Hyposoter exiguae* (larval endoparasitic wasp) *Trichogramma* spp. (egg endoparasitic wasps)	 61 56–60
	citrus cutworm variegated cutworm	*Banchus* sp. (larval endoparasitic wasp) fungus (pathogen) *Ophion* sp. (larval-pupal endoparasitic wasp)	 120 59
	diamondback moth	*Diadegma insularis* (larval-pupal endoparasitic wasp) *Hyposoter* sp. (larval endoparasitic wasp) *Trichogramma* spp. (egg endoparasitic wasps)	 60–61 56–60
	grape leaffolder	*Bracon cushmani* (larval ectoparasitic wasp)	61
	gypsy moth	beetles (predators) *Entomophaga maimaiga* (pathogenic fungus) *Formica* spp. (predatory ants) nuclear polyhedrosis virus (pathogen) wasps, about 2 dozen spp. (parasites)	89 120–123 99–100 122–124 57
	imported cabbageworm	*Apanteles glomeratus* (larval endoparasitic wasp) *Microplitis plutella* (larval endoparasitic wasp) *Pteromalus puparum* (pupal endoparasitic wasp) *Trichogramma* spp. (egg endoparasitic wasps)	61 59 56–60
	Nantucket pine tip moth	*Campoplex frustranae* (larval-pupal endoparasitic wasp)	 59
	navel orangeworm	*Goniozus legneri* (larval ectoparasitic wasp) *Pentalitomastix plethoricus* (egg-pupal endoparasitic wasp)	62 62
	obliquebanded leafroller	*Glypta variegata* (larval endoparasitic wasp) *Macrocentrus iridescens* (larval-pupal endoparasitic wasp)	
	omnivorous leafroller	*Elachertus proteoteratis* (larval ectoparasitic wasp) *Erynnia tortricis* (larval-pupal endoparasitic fly) *Trichogramma* spp. (egg endoparasitic wasps)	 75 56–60

	oriental fruit moth	*Apanteles* sp. (larval endoparasitic wasp)
		Ascogaster quadridentata (egg-prepupal
		  endoparasitic wasp)
		Bracon sp. (larval parasitic wasp)
		Macrocentrus ancylivorus (larval-pupal
		  endoparasitic wasp)
			61
			
			61

CATERPILLARS (continued)	Specific pest	Natural enemies	Page

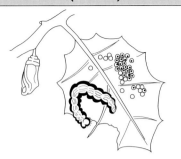

	Specific pest	Natural enemies	Page
	orange tortrix	*Actia interrupta* (larval parasitic fly)	
		Apanteles aristoteliae (larval endoparasitic wasp)	61
		Exochus nigripalpis (larval endoparasitic wasp)	
		Exochus sp. (endoparasitic wasp)	59
		Hormius basalis (parasitic wasp)	
		Hemerobius pacificus (predatory brown lacewing)	103
		Nemorilla pyste (larval-pupal endoparasitic fly)	
	peach twig borer	*Bracon gelechiae* (larval-pupal parasitic wasp)	61
		California gray ant (predator)	99–100
		Erynnia sp. (larval-pupal endoparasitic wasp)	75
		Euderus cushmani (larval endoparasitic wasp)	
		grain mite or itch mite (predator)	
		Macrocentrus ancylivorus (larval parasitic wasp)	
		Paralitomastix varicornis (egg-larval endoparasitic wasp)	
	pink bollworm	*Bracon platynotae* (larval parasitic wasp)	61
		Trichogramma bactrae (egg endoparasitic wasp)	
	redhumped caterpillar	*Apanteles* sp. (larval endoparasitic wasp)	61
		Hyposoter fugitivus (larval endoparasitic wasp)	60–61
	silverspotted tiger moth	*Uramyia halisidotae* (larval-pupal endoparasitic fly)	73
	western grapeleaf skeletonizer	*Apanteles harrisinae* (larval-pupal endoparasitic wasp)	
		Amedoria misella (larval-pupal endoparasitic fly)	
		granulosis virus (pathogen)	123
	western tussock moth	*Dibrachys* sp. (parasitic wasp)	59
		Hyposoter exiguae, H. fugitivus (larval endoparasitic wasps)	60–61
		Trogoderma sternale (predatory beetle)	

FLIES	Specific pest	Natural enemies	Page

	Specific pest	Natural enemies	Page
	fungus gnats	*B. thuringiensis* ssp. *israelensis* (pathogen)	118
		Steinernema feltiae (entomopathogenic nematode)	120
		Hypoaspis miles (predatory mite)	108–109
	Liriomyza spp. leafminers	*Chrysocharis* spp. (larval-pupal endoparasitic wasps)	
		Dacnusa spp. (larval endoparasitic wasps)	72
		Diglyphus spp. (larval ectoparasitic wasps)	72
		Opius sp. (larval-pupal endoparasitic wasp)	
		Solenotus intermedius (parasitic wasp)	
	mosquitoes	*Bacillus sphaericus* (pathogen)	118
		B. thuringiensis ssp. *israelensis* (pathogen)	118
		beetles, aquatic (predators)	80
		bugs, aquatic (predators)	91
		Lagenidium giganteum (pathogen)	
		mosquito fish (predator)	

Natural Enemies attacking many flies	Page
beetles (predators)	80
microorganisms (pathogens)	117
spiders (predators)	110
wasps (parasites)	57
wasps (predators)	98

FLIES (*continued*)

Specific pest	Natural enemies	Page
tentiform leafminers, *Phyllonorycter* spp.	*Pnigalio flavipes* (larval ectoparasitic wasp)	8
	Sympiesis stigmata (larval ectoparasitic wasp)	8
	lacewings (predators)	100
	lady beetle larvae (predators)	80

GRASSHOPPERS, CRICKETS

Natural enemies	Page
blister beetles (predators)	
flies (parasites)	73
Nosema locustae protozoan (pathogen)	
wasps (parasites)	57

LEAFHOPPERS

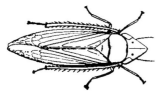

Specific pest	Natural enemies	Page
grape leafhopper	*Anagrus* spp. (egg endoparasitic wasps)	58
	spiders (predators)	110

Natural Enemies attacking many leafhoppers	Page
bugs (predators)	91
flies (parasites)	73
lacewings (predators)	100
mites (predators)	106
spiders (predators)	110
wasps (parasites)	57

LEAFMINERS (*see* FLIES)

MEALYBUGS

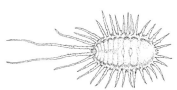

Natural Enemies attacking many mealybugs	Page
lacewings (predators)	100
mealybug destroyer lady beetle (predator)	87
midge larvae (predators)	95
minute pirate bugs (predators)	91
wasps (parasites)	57

Specific pest	Natural enemies	Page
citrophilus mealybug	*Coccophagus gurneyi* (nymphal endoparasitic wasp)	71
	Hungariella (=Tetracnemoides) pretiosa (nymphal endoparasitic wasp)	
citrus mealybug	*Leptomastix dactylopii* (nymphal-adult endoparasitic wasp)	70
	Leptomastidea abnormis (nymphal endoparasitic wasp)	70
	mealybug destroyer (predatory lady beetle)	87
Comstock mealybug	*Allotropa burrelli* (nymphal-adult endoparasitic wasp)	71
	A. convexifrons (nymphal endoparasitic wasp)	
	Pseudaphycus malinus (nymphal-adult endoparasitic wasp)	
grape mealybug	*Acerophagus notativentris* (nymphal-adult endoparasitic wasp)	
longtailed mealybug	*Anarhopus sydneyensis* (nymphal-adult endoparasitic wasp)	71

MITES

Natural Enemies attacking many mites	Page
bigeyed bugs (predators)	91
dustywings (predators)	103
lacewings (predators)	100
midge larvae (predators)	95
minute pirate bugs (predators)	91
mites (predator)	106
sixspotted thrips (predator)	106
spider mite destroyer lady beetle (predator)	88
western flower thrips (predator)	106

Specific pest	Natural enemies	Page
broad mites	*Neoseiulus* spp. (predatory mites)	108
brown mite	brown lacewings (predators)	103
	western predatory mite (predator)	107
citrus red mite	*Conwentzia barretti* (predatory dustywing)	103
	Euseius stipulatus (predatory mite)	
	Euseius tularensis (predatory mite)	107
	viral disease (pathogen)	124
cyclamen mite	sixspotted thrips (predator)	106
	Typhlodromus spp. (predatory mites)	106
Tetranychus spp. European red mite	*Amblyseius (=Neoseiulus)* spp. (predatory mites)	108
	bigeyed bugs (predators)	93
	damsel bugs (predators)	94
	lacewings (predators)	100
	Phytoseiulus persimilis (predatory mite)	107
	sixspotted thrips (predator)	106
	spider mite destroyer (predatory lady beetle)	88
	western predatory mite (predator)	107

NATURAL ENEMIES

NEMATODES, PLANT PARASITIC

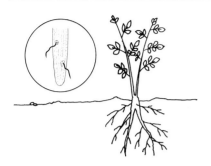

	Natural enemies	Page
	nematodes (predators)	34
	microorganisms (parasites and predators)	34
	suppressive crops and plant by-products (competitors and antagonists)	33

PLANT PATHOGENS

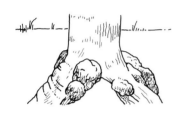

Natural Enemies attacking plant pathogens	Page
bacteria and fungi	28
mycopesticides	29
mycorrhizae	31
suppressive soils	30

(antagonists, competitors, parasites, and predators, sometimes in combination; type of action poorly understood for some natural enemies)

Specific pest	Natural enemies	Page
crown gall	*Agrobacterium radiobacter* K84 (bacterium)	28
Annosus root disease of pines	*Phanerochaete (=Peniophora) gigantea* (fungus)	29
fireblight and frost of pears	*Pseudomonas cepacia, P. fluorescens* (bacteria)	28
soilborne root and crown decays caused by *Fusarium, Pythium,* and *Rhizoctonia* spp.	*Bacillus, Burkholderia, Gliocladium, Pseudomonas, Streptomyces,* and *Trichoderma* spp. (beneficial bacteria and fungi)	29

PSYLLIDS

Natural Enemies attacking many psyllids	Page
beetles (predators)	80
bugs (predators)	91
lacewings (predators)	100
minute pirate bugs (predators)	91
wasps (parasites)	57

Specific pest	Natural enemies	Page
acacia psyllid	*Anthocoris nemoralis* (predatory minute pirate bug)	91
	Diomus pumilio (predatory lady beetle)	71
blue gum psyllid	*Psyllaephagus pilosus* (nymphal endoparasitic wasp)	70
eugenia psyllid	*Tamarixia* sp. (nymphal endoparasitic wasp)	70
pear psylla	lacewings (predators)	100
	Leucopis spp. (predatory aphid flies)	95
	minute pirate bugs (predators)	91
peppertree psyllid	*Tamarixia* sp. (nymphal endoparasitic wasp)	71
potato psyllid	damsel bug (predator)	94
	minute pirate bug (predator)	91

SAWFLIES, PEARSLUG	Natural enemies	Page
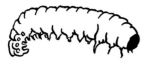	beetles (predators)	80
	birds and small mammals (predators)	116
	flies (parasites)	73
	viruses (pathogens)	123
	wasps (parasites)	57

SCALES

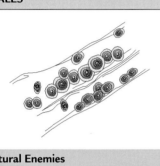

Natural Enemies attacking many scales	Page
beetles (predators)	85
lacewings (predators)	100
minute pirate bugs (predators)	91
mites (predators)	106
wasps (parasites)	64

Specific pest	Natural enemies	Page
black scale	*Hyperaspis quadrioculata* (predatory lady beetle)	85–86
	Metaphycus bartletti (nymphal-adult endoparasitic wasp)	
	Metaphycus helvolus (nymphal endoparasitic wasp)	65
	Scutellista caerulea (=*S. cyanea*) (adult ectoparasitic and egg predatory wasp)	
brown soft scale	*Metaphycus* spp. (nymphal endoparasitic wasps)	64–65
	Microterys nietneri (=*M. flavus*) (nymphal endoparasitic wasp)	64–65
	Chilocorus spp. (predatory lady beetles)	85–87
	Rhizobius lophanthae (predatory lady beetle)	85
California red scale	*Aphytis melinus, Aphytis* spp. (nymphal-adult ectoparasitic wasps)	65
	Comperiella bifasciata (nymphal-adult endoparasitic wasp)	65–66
	Chilocorus spp. (predatory lady beetles)	85–87
	Rhizobius lophanthae (predatory lady beetle)	85
citricola scale	*Coccophagus lycimnia* (nymphal-adult endoparasitic wasp)	58
	Coccophagus scutellaris (nymphal-adult endoparasitic wasp)	
	Metaphycus flavus (nymphal endoparasitic wasp)	
	Metaphycus luteolus (nymphal endoparasitic wasp)	64
cottony cushion scale	*Cryptochaetum iceryae* (nymphal-adult endoparasitic fly)	77
	vedalia beetle (predatory lady beetle)	85
dictyospermum scale	*Aphytis melinus* (nymphal-adult ectoparasitic wasp)	65
European elm scale	*Trichomasthus coeruleus* (nymphal-adult endoparasitic wasp)	

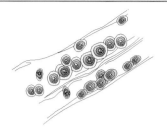

Specific pest	Natural enemies	Page
European fruit lecanium frosted scale	*Aphytis* spp. (nymphal-adult ectoparasitic wasps)	64
	Coccophagus lecanii (nymphal-adult endoparasitic wasp)	58
	Encarsia spp. (nymphal endoparasitic wasps)	
	Metaphycus spp. (nymphal endoparasitic wasps)	64–65
	Metaphycus californicus (nymphal endoparasitic wasp)	8
fig scale	*Aphytis mytilaspidis* (nymphal-adult ectoparasitic wasp)	64
green shield scale	mealybug destroyer (predatory lady beetle)	87
	Metaphycus spp. (endoparasitic wasps)	64–65
hemispherical scale	*Metaphycus* spp.	
	Metaphycus helvolus (nymphal endoparasitic wasps)	64–65
ice plant scales	*Metaphycus funicularis* (nymphal-adult endoparasitic wasp)	64
	Metaphycus stramineus (nymphal-adult endoparasitic wasp)	
	Exochomus flavipes (predatory lady beetle)	
nigra scale	*Metaphycus helvolus* (nymphal endoparasitic wasp)	64–65
obscure scale	*Encarsia aurantii* (nymphal endoparasitic wasp)	
oleander scale	*Aphytis* spp. (nymphal-adult ectoparasitic wasps)	64–66
olive scale	*Aphytis maculicornis* (nymphal-adult ectoparasitic wasp)	64–65
	Coccophagoides utilis (nymphal-adult endoparasitic wasp)	
purple scale	*Aphytis lepidosaphes* (nymphal-adult ectoparasitic wasp)	64–65
	Chilocorus spp. (predatory lady beetles)	85–87
	Rhyzobius lophanthae (predatory lady beetle)	85
San Jose scale	*Aphytis* spp. (nymphal-adult ectoparasitic wasps)	64–66
	Chilocorus orbus (predatory lady beetle)	85–87
	Cybocephalus californicus (predatory sap beetle)	90
	Encarsia perniciosi (endoparasitic parasitic wasp)	
walnut scale	*Aphytis* spp. (ectoparasitic wasps)	64–66
	Chilocorus orbus (predatory lady beetle)	85–87
	Cybocephalus californicus (predatory sap beetle)	90
	Encarsia (=*Prospaltella*) spp. (nymphal endoparasitic wasps)	
yellow scale	*Comperiella bifasciata* (endoparasitic wasp)	65–66

SNAILS, SLUGS

Natural Enemies attacking many snails and slugs	Page
flies (parasites)	74
Ocypus spp. rove beetles (predators)	90
Scaphinotus spp. ground beetles (predators)	89
vertebrates (predators)	116

Specific pest	Natural enemies	Page
brown garden snail	decollate snail (predator)	115

THRIPS

Natural Enemies attacking many thrips	Page
minute pirate bugs (predators)	91
mites (predators)	106
thrips (predators)	106
wasps (parasites)	57

Specific pest	Natural enemies	Page
citrus thrips	*Anystis agilis* (predatory mite)	108
	Euseius tularensis (predatory mite)	107
	minute pirate bug (predator)	91
Cuban laurel thrips	lacewings (predators)	100
	Macrotracheliella nigra (predatory minute pirate bug)	
greenhouse thrips	*Franklinothrips vespiformis* (predatory thrips)	106
	Thripobius semiluteus (nymphal endoparasitic wasp)	72

WEEDS

Natural Enemies attacking many weeds	Page
fungi (pathogens)	37
invertebrates and vertebrates (herbivores and seed predators)	36
plants (allelopaths and competitors)	35

Specific pest	Natural enemies	Page
Klamath weed or St. Johnswort	Klamathweed beetle (herbivorous leaf beetle)	38–39
northern jointvetch (in rice and soybeans in eastern U.S.)	*Colletotrichum gloeosporioides* (mycoherbicide)	36–37
prickly pear cacti	cactus moth (cactus boring larvae)	40
	cochineal insect (sucking insect)	40
	microorganisms (plant pathogens)	
puncturevine	*Microlarinus lypriformis* (stem mining weevil)	39–40
	Microlarinus lareynii (seed weevil)	36
rush skeletonweed	*Puccinia chondrillina* (mycoherbicide rust fungus)	37
stranglervine (in Florida citrus orchards)	*Phytophthora palmivora* (mycoherbicide fungus)	36–37

Specific pest	Natural enemies	Page
thistles, various spp.	flies, moths, and weevils (seed-, stem-, and root-feeders)	36–40
various forbs, grasses, and shrubs	domesticated livestock (herbivores)	36, 38

WEEVILS (*see* BEETLES)

WHITEFLIES

Natural Enemies attacking many whiteflies	Page
bigeyed bugs (predators)	91
lacewings (predators)	100
lady beetles (predators)	87
wasps (parasites)	68

Specific pest	Natural enemies	Page
ash whitefly	*Clitostethus arcuatus* (predatory lady beetle)	71
	Encarsia inaron (nymphal-pupal endoparasitic wasp)	68
bayberry whitefly	*Encarsia* spp. (endoparasitic wasps)	68
	Eretmocerus spp. (ecto- and endoparasitic wasps)	71
citrus blackfly	*Amitus hesperidum* (nymphal-pupal endoparasitic wasp)	71
	Encarsia opulenta (nymphal-pupal endoparasitic wasp)	68
citrus whitefly	*Encarsia* spp. (endoparasitic wasps)	68
	Eretmocerus spp. (ecto- and endoparasitic wasps)	71
greenhouse whitefly	*Encarsia formosa* (nymphal-pupal endoparasitic wasp)	68–69
silverleaf whitefly	*Encarsia* spp. (endo- and ectoparasitic wasps)	68–69
	Eretmocerus spp. (endo- and ectoparasitic wasps)	
	Delphastus pusillus (predatory lady beetle)	87
woolly whitefly	*Amitus spiniferus* (parasitic wasp)	71
	Cales noacki (nymphal endoparasitic wasp)	
	Eretmocerus sp. (ecto- and endoparasitic wasp)	

Note: Consult the index at the end of this book for complete listings and common and scientific names of the organisms mentioned here. Endoparasitic species feed inside their host; ectoparasites feed attached to their host's outside. The beetle illustation is from Anonymous (1960); the leafhopper is from Anonymous (1952); the mosquito is from Cole (1969), reprinted with permission from University of California Press; the mealybug by P. J. Hollyoak and grasshopper are from Gorham (1991).

CHAPTER THREE

NATURAL ENEMIES OF PLANT PATHOGENS

D ISEASE-CAUSING microorganisms can injure or kill plants. These plant pathogens include certain species of bacteria, fungi, viruses, and phytoplasmas. Environmental stress or abiotic disorders, insects, nematodes, and other pests sometimes cause damage that may be confused with symptoms caused by plant pathogens. These agents may also act in conjunction with pathogens to cause disease. Because many diseases cannot be cured after pathogens infect plants, control strategies (whether chemical, biological, or cultural) focus on excluding, eradicating, or inhibiting germination and penetration by propagules (such as spores and other structures that disperse and initiate new infections) or increasing the ability of plants to resist infection.

Many naturally occurring microorganisms kill or retard growth of plant pathogens (Lumsden, Lewis, and Locke 1993; O'Neill et al. 1996; Swadling and Jeffries 1996). However, little is known about most of the biological control agents involved, and they generally cannot be identified in the field by pest managers. In some cases, soil solarization (Elmore et al. 1997), crop rotation,

or incorporating compost, green manure, or other amendments can increase activities of beneficial soil organisms and reduce disease occurrence. Solarization has been used to increase fluorescent pseudomonad and *Bacillus* spp. bacteria known to be natural enemies of soilborne pathogens (Stapleton and DeVay 1984), for example, in the root zone of lettuce (Gamliel and Stapleton 1993). However, relatively few specific recommendations for using these techniques in practical management situations can be made at this time.

Some naturally occurring disease-suppressive soils are recognized, but the mechanisms of disease suppression by these soils are not well understood. However, soil organisms are at least partially involved because suppression can be destroyed by chemical fumigation or heat pasteurization.

Certain fungi and bacteria called mycopesticides are commercially available for disease control. Other than disease-suppressive composts and amendments, mycopesticides are currently the main strategy growers can use to implement biological control of plant pathogens. Most biological control agents currently available for practical use target soilborne pathogens. However, there is potential for commercial agents for use against foliar diseases. The competitive bacterium *Pseudomonas fluorescens* A506 has been used effectively in combination with antibiotics to control fireblight disease of pears in commercial pear orchards in California (Lindow, McGourty, and Elkins 1996). Honey bees inoculated with *P. fluorescens* and *Erwinia herbicola*, another beneficial bacterium, have been used to disseminate these fireblight bio-control agents while bees are pollinating apple and pear orchards (Thomson et al. 1992). Because temperature and moisture are key factors contributing to the development of many diseases, the ability to control environmental conditions in greenhouses makes biological control of foliar diseases a promising option in greenhouses (Jarvis 1992, Marois 1992).

Steam is often used to treat potting soil in nurseries. Heat kills most pests that infest media, but some pathogens can soon recolonize pasteurized media. Acting as sapro-phytes feeding on the available organic matter in the absence of nonpathogenic microorganisms, these pathogens can quickly build up to large populations. Applying certain nonpathogenic microorganisms to media after pasteurization allows these bene-ficial microorganisms to get a head start in colonization; through competition, exclu-sion, and antibiosis, these nonpathogens may prevent later-arriving root-decay pathogens from becoming abundant.

MECHANISMS OF PLANT PATHOGEN BIOLOGICAL CONTROL

Natural enemies of plant pathogens may exert control through predation, parasitization, competition, or antibiosis (Whipps 1992). Induced resistance (Tuzun and Kloepper 1995), a complex mechanism of preventing pathogen infection that can be similar to vaccinating people by exposing them to non-pathogenic forms of the disease-causing agent, is not discussed further here. Competition and antibiosis probably hold the most promise for practical management of plant disease. Some biological control agents exert control through two or more of these mechanisms; for many agents the exact control mechanism is not well understood.

■ **Competition.** Some biological control agents consume the same nutrients and other limited resources or colonize the same spaces as plant pathogens. If competitor species that do not injure living plants are applied or occur naturally before the pathogen population is

high, these competitors can exclude or limit development of disease-causing organisms. Certain commercially available biological control agents, including *Gliocladium*, *Pseudomonas*, and *Tricho-derma* spp., can be applied to clean seed or media after pasteurization. Through competition and exclusion, these non-pathogens can reduce or prevent damage from *Fusarium oxysporum*, *Pythium* spp., and *Rhizoctonia solani* (Cartwright and Benson 1995; Datnoff, Nemec, and Pernezny 1995; Isakeit et al. 1991, 1993; Lewis, Lumsden, and Locke 1996). For example, avocado and citrus seedling growth was enhanced by certain yard waste (wood chips, grass, and leaves) or rice hull mulches that had been fumigated and then inoculated with *Trichoderma harzianum* and *Pseudomonas fluorescens* (Casale et al. 1995). *Bacillus subtilis*, *Gliocladium virens*, and *Pseudomonas* and *Tricho-derma* spp. are applied as vegetable crop seed inoculants to prevent damping-off from *Pythium* and *Rhizoctonia* in the greenhouse and field (Fukui et al. 1994a, 1994b; Zhang, Howell, and Starr 1996).

■ **Antibiosis.** Some microorganisms produce compounds that kill or sup-press populations of certain other microorganisms. Streptomycin, a by-

This crown gall (*Agrobacterium tumefaciens*) growing around the tree base might have been prevented by *Agrobacterium radiobac-ter* strain K84, a microorganism that pro-duces an antibiotic that is specific against other *Agrobacterium* bacteria that cause plant galls. This beneficial microorganism is widely used to protect fruit trees from infection by dipping roots or cuttings into a protective suspension of commercially available *A. radiobacter* strain K84 for 30 seconds before planting or immediately after wounding trees. However, infection by some *A. tumefaciens* strains, such as that affecting grapes, is not prevented by *A. radiobacter* strain K84.

Inoculating freshly cut pine stumps with *Phanerochaete* or *Peniophora gigantea* can provide commercial control of Annosus root disease (*Heterobasidion annosum*). This antagonistic fungus prevents the pathogen from entering stumps and spreading through roots to nearby healthy pines. *Phanerochaete gigantea* can be applied by adding the fungal spores to chainsaw blade lubricating oil before making cuts.

Properly applying clear plastic to bare soil for several weeks during sufficiently warm, sunny weather superheats the soil and can control pathogens, nematodes, and certain weeds near the surface (Elmore et al. 1997). Soil solarization also favors the survival and reproduction of heat-tolerant beneficial microorganisms that are antagonists to soilborne pathogens. Pest control can sometimes be increased by solarization in combination with incorporation of crop residues (Stapleton and DeVay 1995).

product of soil-dwelling *Streptomyces griseus* bacteria, is used as a seed dip to prevent damping-off. This antibiotic is also used as a cutting dip or foliar spray to control bacterial leaf, stem, or wilt diseases. Certain *Bacillus, Penicillium, Pseudomonas,* and *Streptomyces* spp. are used as root dips or seed inoculants to produce antibiotics that prevent damping-off caused by *Phytophthora, Pythium,* and *Rhizoctonia* fungi (Fang and Tsao 1995).

■ **Parasitism and Predation.** Predation and parasitism are often used in the biological control of arthropods pests. Predation and parasitism of microorganisms is less well known. Certain nematodes and amoebae prey on bacteria and fungi. Some microorganisms parasitize other microbes, a phenomenon called hyperparasitism (parasitism of one parasite by another). Although parasites and predators may be components of certain disease-suppressive soils, currently there is little information on the practical use of parasitism and predation for pathogen control.

■

MYCOPESTICIDES

Mycopesticides are commercially available beneficial microorganisms or their by-products that control plant pathogens (table 3-1). They are also

called mycofungicides because the most common products target pathogenic fungi. Most mycopesticides only prevent infection, such as those added to pasteurized media or clean seeds, but some (*Streptomyces* spp.) have curative action. Mycopesticides must be registered and labeled in accordance with pesticide regulations. Because of these regulations, the testing requirements and available information on mycopesticides may be more extensive than information on microbes available in largely unregulated amendments or inoculants.

■

DISEASE-SUPPRESSIVE COMPOST AND SOIL AMENDMENTS

Composting is the biological decomposition of organic material (such as tree bark) under controlled conditions. Heat generated by decomposer microorganisms during composting can destroy most pathogens, weed seeds, and invertebrates. Beneficial microorganisms inhabiting some composts can control pathogenic *Fusarium, Pythium,* and *Rhizoctonia* spp. (Hoitink and Gre-

bus 1994, Quarles and Grossman 1995). Disease-suppressive compost is used on a large scale in some commercial nurseries. High-quality compost also can improve nutritional, physical, and chemical characteristics of media, enhancing crop growth.

Composting requires careful monitoring and management using controlled conditions to produce sufficiently decomposed, stable material that consistently benefits plant growth. Heat naturally generated by decomposer microorganisms must reach temperatures of 140°F (60°C) or more for at least 3 days for heat to sufficiently destroy pathogens and most other pests. Regular mixing to ensure that materials on the edge are moved into the pile where they can reach adequate treatment temperatures is one of several activities needed to produce good compost. Disease-suppressive microorganisms may naturally recolonize compost during a final curing phase (fig. 3-1); compost properly cured for 4 or more months is often disease suppressive. However, inoculating compost during the curing phase with biological control agents may be

TABLE 3-1. Selected commercially available mycopesticides.

MICROORGANISMS (TRADE NAMES)	PATHOGENS CONTROLLED	APPLICATION METHODS
Agrobacterium radiobacter strain K84 (Galltrol-A)	crown gall (*Agrobacterium tumefaciens*)	preplant preventive as cutting, root, or seed dip
Ampelomyces quisqualis (AQ-10)	powdery mildew	preventive foliar spray
Bacillus subtilis (Kodiak)	damping-off fungi, such as *Pythium*	seed inoculant
Burkholderia cepacia (Deny)	*Fusarium* and *Rhizoctonia* root rots and certain nematodes	preventive application prior to planting as cutting, seed, or seedling dip
Candida oleophila (Aspire)	postharvest fruit decay	preventive application to harvested fruit
Gliocladium virens GL-21 (GlioGard, SoilGard)	*Pythium* and *Rhizoctonia*	incorporation in soil or media and incubation before planting
Pseudomonas cepacia, P. fluorescens (BlightBan A506)	fireblight and frost of pears	foliar spray
Pseudomonas syringae (BioSave)	postharvest fruit decay	postharvest preventive application to certain fruits before storage
Streptomyces griseoviridis Strain K61 (Mycostop)	root decay fungi, e.g., *Alternaria, Fusarium, Phomopsis*	preventive as dip, drench, or spray for seeds and container-grown plants
Streptomycin from *Streptomyces griseus* (Agrimycin 17)	bacterial blights, cankers, leaf spots, and wilts; crown gall	curative as cutting dip or foliar spray
Trichoderma harzianum, T. polysporum (Binab, Bio-Trek, F-Stop, RootShield)	*Pythium* and other soilborne fungi	seed and bulb dip, soil drench, tree wound dressing

Note: Mycopesticides must be labeled and registered in accordance with pesticide regulations of the U.S. Environmental Protection Agency (and in California with the Department of Pesticide Regulation). Check labels for permitted uses and methods. Many of these microorganisms are also available in commercial amendments or inoculants, which are largely unregulated, and information on their use may be less reliable than the information for registered mycopesticides.

Sources: Cook et al. 1996, Quarles 1996b.

needed to ensure consistent disease suppression, to provide specific suppression of certain pathogens, and to reduce curing time.

The disease-suppressive ability of compost is influenced by many physical, chemical, and biological factors during the composting process and while compost is being used. Unfavorable properties or conditions can negate the beneficial effects of compost. Although proper composting eliminates most pathogens, unpasteurized soils and organic materials can be contaminated by pathogens.

The disease-suppressive characteristics of composts can be variable; quality control is a great impediment to consistently successful use of compost for disease suppression. Inadequate decomposition may be the most common problem. Unlike mycopesticides, composts and amendments are largely unregulated. Effective disease suppression using compost requires adequate knowledge and properly prepared material. See *Compost Production and Utilization: A Growers' Guide* (van Horn 1995) and other sources (such as Hoitink and Grebus 1994) for details.

Certain natural toxins produced by plants and animal by-products can also inhibit some pathogens in a manner similar to allelopathy as discussed for weed control. For example, *Fusarium oxysporum* development in cabbage can sometimes be reduced by incorporating residues of certain crops (such as cole crops) into soil, then allowing the residues to naturally decompose or be heated through soil solarization before planting. Similarly, Gamliel and Stapleton (1993) found that a root knot nematode (*Meloidogyne incognita*) in lettuce was completely controlled by combining composted chicken manure with solarization when neither treatment alone was effective.

SUPPRESSIVE SOILS

Some naturally occurring soils and certain peats improve plant growth despite the presence of certain plant pathogens (Schroth and Hancock 1982). Natural suppression is associated with beneficial microorganisms and certain physical and chemical characteristics that affect the soil microbiology. Induced suppression is somewhat independent of soil characteristics and largely results from certain cropping and cultural practices. A variety of biological, chemical, and physical properties can produce effective soil suppressiveness.

Suppression of wilt diseases caused by strains of *Fusarium oxysporum* is a common example of natural suppres-

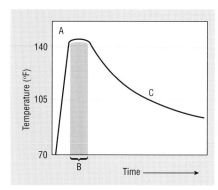

Before conifers such as these are planted in the field for reforestation, seedlings in nurseries are often inoculated with beneficial mycorrhizal fungi to improve plant survival and growth.

■ FIGURE 3-1. The three phases of composting: A: an initial few days as temperatures quickly rise to 105° to 120°F (41° to 50°C) as easily degraded substances decompose. B: a minimum of 3 days during which heat generated by decomposer microorganisms must reach at least 140°F (60°C) to kill most plant pathogens, weed seeds, and invertebrate pests. C: a curing phase when temperatures decline and stabilize at just under 105°F (41°C) and microorganisms naturally recolonize or are inoculated into the material. Proper composting can produce disease-suppressive media for growing plants. However, disease suppression by compost is influenced by many physical, chemical, and biological factors during the composting process and when compost is being used.

This equipment screens ground up tree trimmings and yard waste before sending the material to be composted in rows seen in the background. Production of good compost requires careful monitoring and good management, including adequate aeration, regular mixing, and sufficient moisture. Compost used in container media helps to control soilborne pathogens, but effective disease suppression using compost requires adequate knowledge and properly prepared material.

sion. For example, Fusarium wilt of carnation is greatly reduced or avoided when carnations are planted in some field soils in the Salinas Valley, California (Isakeit et al. 1991). This suppression, which can be related to the presence of *Pseudomonas* spp. or other beneficial microorganisms, may also be at least partly dependent on certain types of clay and soil minerals.

The reasons why some natural soils suppress disease are complex and often poorly understood. Suppressive soils can be ineffective when moved to a different location. Unless growers are lucky enough to be planting in suppressive soils, use of mycopesticides or on-site production of disease suppressive media would be more practical approaches to biological control of pathogens in many situations.

■

MYCORRHIZAE

Soil-dwelling fungi called mycorrhizae form beneficial associations with the roots of most flowering plants. Mycorrhizae can make roots more resistant to infection by certain fungi (such as *Fusarium, Phytophthora,* and *Pythium*), improve plants' ability to absorb nutrients, and may aid in water uptake (Calvet, Pera, and Barea 1993; Pfleger and Linderman 1994;

Zak and Ho 1994). Endomycorrhizae may form a loose mycelial growth on roots, which may not be visible to the naked eye. Mycelia sometimes produce pearl-like reproductive structures, which may be visible. Ectomycorrhizae can be black, white, or colorful and commonly cause roots to appear swollen, forked, or galled. The appearance of ectomycorrhizae may be confused with other causes of root galls, such as certain invertebrate pests or the whitish galls of beneficial nitrogen-fixing bacteria.

Mycorrhizae improve cotton growth (Davis et al. 1996, Watkins 1981), and certain crop rotations can improve this mycorrhizal activity. Safflower roots are more heavily colonized by *Glomus fasciculatus* mycorrhizae than most other crops. Cotton planted following safflower becomes more heavily colonized than normal with this beneficial mycorrhizal fungus.

Mycorrhizal fungi can be isolated from soil containing fungi known to be beneficial. Some nurseries producing avocado, citrus, and other stock inoculate their plants with mycorrhizae. The most study of mycorrhizal interactions has been with forest trees, and most of the deliberate use of mycorrhizae has been in the production of conifer seedlings (Castellano and Molina 1989).

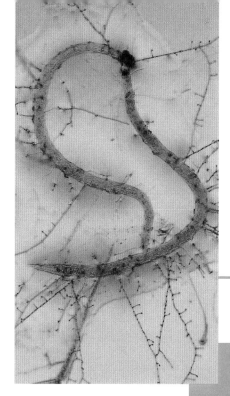

CHAPTER FOUR

NATURAL ENEMIES OF NEMATODES

Magnifying glass signifies that subject is less than 1 mm.

Nematodes are tiny roundworms that feed on plants, animals, and microorganisms. Plant-parasitic nematodes can stunt crop growth, facilitate damage by other pests, and sometimes kill plants. Many other nematode species are innocuous or beneficial to plant production. For example, entomopathogenic nematodes can be used as beneficial species when applied to control certain insect pests.

Plant-parasitic nematodes traditionally have been controlled through sanitation (such as preventing introductions, removing infested plants, pasteurizing media, cleaning tools), cultural practices (such as water management, crop rotation, fallowing, solarization, organic amendments), planting resistant or nonhost crops and nursery-certified nematode-free stock, and soil-applied nematicides.

Although biological control agents reduce plant pathogenic nematodes in certain situations, biological control has not been developed as a practical method of nematode control. A limiting factor is the inability to consistently transfer biological benefits from one site to another. Biological control is complicated by poorly understood physical, chemical, and biological interactions in soil (Ferris et al. 1992,

Sterling 1991). Research in molecular biology is improving the ability to identify nematodes and understand their ecology, facilitating development of biologically based controls.

Natural enemies of nematodes may reduce their numbers through one or more mechanisms including antibiosis, competition, parasitism, and predation. Antibiosis is the production by one organism of compounds (such as ammonia, avermectins, and certain fatty acids) that kill, inhibit, or retard development of another organism. Competition from other microorganisms for necessary resources (such as food and space) also suppresses nematode populations.

SUPPRESSIVE CROPS AND PLANT BY-PRODUCTS

Barley, marigold, perennial rye, and certain legumes such as clover and vetch can reduce soil populations of certain plant-parasitic nematodes. These plants produce chemicals (sometimes called allelochemicals) that kill or repel nematodes, suppress their growth, stimulate premature egg hatching, or disrupt the attraction between nematodes seeking to mate. In the future, suppressive plants might be used in crop rotations, as cover crops, in intercropping, and trap crops to

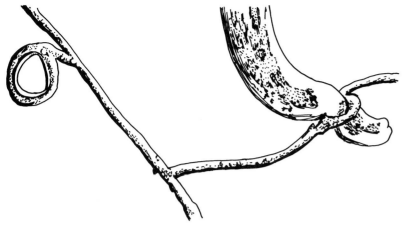

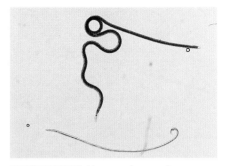

 Nematodes are tiny roundworms, like this male (bottom) and female *Heleidomeris magnapapula*. Not all nematodes are pests; this species is a beneficial internal parasite of adult biting midges.

■ FIGURE 4-1. A nematode at right snared by a ring trap fungus. Another trap waits for prey at the upper left. Certain fungi (such as *Arthrobotrys brochopaga* and *A. dactyloides*) produce these looplike structures in spaces among soil particles where nematodes commonly travel. When a nematode enters, the loop contracts like a noose. Fungal hyphae then grow into the captured nematode and consume its body. Other nematode-trapping fungi use sticky nets (*A. oligospora*), sticky knobs (such as *Dactylaria haptotyla* and *Nematoctonus* spp.), or sticky spores (*Drechmeria coniospora* and *Hirsutella rhossiliensis*).

A heavy root knot nematode (*Meloidogyne* spp.) infestation caused these galls. Root knot nematodes are the most common nematode pests in California.

reduce nematode numbers (Bridge 1996). Although rotating marigolds with crops such as lilies grown for bulb production has been somewhat successful, many impediments to use of suppressive plants have been identified. These include phytotoxicity (seen in many crops when they are planted in rotation after marigolds), inadequate control, and reduced crop sales from land devoted to growing plants that are not marketed. Additionally, nematodes already feeding and reproducing within roots are often unaffected by specific control agents such as allelochemicals. Similarly, most plants are effective only against some nematode species, while other nematodes may use those plants as hosts.

■

PARASITES AND PREDATORS

Various other invertebrates and microorganisms attack nematodes. Certain amoebae, flatworms, mites, protozoa, and springtails parasitize nematodes. Nematode-trapping fungi are unusual predators (Jaffee, Muldoon and Tedford 1992; Jaffee and Muldoon 1995) (fig. 4-1). One example of applied nematode biological control is the maintenance of monocultures of small grains in England to support high

numbers of naturally occurring fungi that control cereal cyst nematode.

Predaceous nematodes (such as *Eudorylaimus*, *Labronema* and *Seinura* spp.) pierce plant-feeding nematodes with a hollow stylet and suck out the pests' body contents. *Mononchus* and *Odontopharynx* spp. nematodes slit nematodes' bodies using a large tooth, then ingest the exuding contents. Some predaceous nematodes swallow prey nematodes whole. Although predaceous nematodes may be beneficial in perennial crops with relatively undisturbed soils, their practical value is largely unknown. Predaceous nematodes are susceptible to soil disturbance and are uncommon in cultivated soils even when they are abundant in adjacent untilled soils.

Bacteria (Sikora and Hoffmann-Hergarten 1993) and fungi (Meyer and Huettel 1993) are promising biological control agents. These potential microbial agents include *Pasteuria penetrans* bacteria, which have spores that adhere to and infect nematodes, and *Catenaria* spp. fungi, which have mobile zoospores that swim through moist soil to find and infect nematodes. However, mass production of pathogens is difficult, and economical and effective field application methods are yet to be developed.

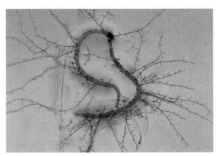

 The fungus *Hirsutella rhossiliensis* has killed this cyst nematode (*Heterodera* sp.). The fungal hyphae strands radiating from this dead soil-dwelling nematode develop protruding structures that produce sticky spores that attach to and infect passing nematodes (Jaffee, Muldoon, and Westerdahl 1996; Tedford, Jaffee, and Muldoon 1995).

CHAPTER FIVE

NATURAL ENEMIES OF WEEDS

WEEDS ARE PLANTS growing where people find them to be undesirable. Many methods are used to control weeds. Cultivation, by hand and more recently with machines, has been a major management method for centuries. Herbicides continue to increase in importance, with more herbicides applied than all other groups of pesticides combined. Mulches, landscape fabrics, good irrigation management, proper sanitation practices, and biological control are also used, often in combination with cultivation and herbicides. Biological weed control has been most successful for specific problem weeds on noncropped lands such as forests, rangelands, and roadsides. In cultivated crops, especially in the western United States, biological control of weeds has not yet had much impact.

■

MECHANISMS OF WEED BIOLOGICAL CONTROL

Biological weed control relies primarily on: competition with other plants (including antagonism or allelopathy), pathogenicity, and herbivory (Turner et al. 1992). Herbivores (invertebrates and vertebrates) and pathogens (microorganisms) can kill weeds, reduce weed reproduction and dispersal, or stress weeds so that more resources are available for use by desirable, competing plant species.

Many severe weeds are of foreign origin. Because these plants were introduced without their natural enemies, much research has focused on classical or inoculative introduction of natural enemies (Harley and Forno 1992, Julien 1992, Nechols et al. 1995). Invertebrates and pathogens chosen for these types of biological weed control programs are usually species that attack only one or a few closely related weedy plant species (Blossey 1995) and have the ability to significantly control the target pest (table 5-1).

■ **Competition and Allelopathy.** Plants growing with other plants compete for limited resources. If desirable or innocuous plants can monopolize light, water, nutrients, or growing space, they will out-compete and eventually displace weeds. Planting and properly caring for crops and landscapes facilitates desirable plant growth in habitats that might otherwise be occupied by weeds. Cover crops and smother crops can be grown to occupy space and retard weed germination and establishment. Competition between the harvested crop and the smother crop is minimized by good plant selection (such as annual cover

35

TABLE 5-1. Selected biological control agents of weeds.

WEED		BIOLOGICAL CONTROL AGENT		
Common name	*Scientific name*	*Common name*	*Scientific name*	*References*
aquatic weeds	various spp.	grass carp or white amur*	*Ctenopharyngodon idella**	Chilton and Muoneke 1992, Santha et al. 1994
Canadian thistle	*Cirsium arvense*	stem-mining weevil seed weevil gall fly	*Ceutorhynchus litura* *Larinus planus* *Urophora cardui*	Piper and Andres 1995
diffuse knapweed spotted knapweed	*Centaurea diffusa* *Centaurea maculosa*	seedhead gall fly seedhead gall fly root-boring beetle	*Urophora affinis* *Urophora quadrifasciata* *Sphenoptera jugoslavica*	Mays and Kok 1996, Piper and Rosenthal 1995, Story 1995
forbs, grasses, and shrubs	various spp.	cattle	*Bos* spp.	Lanini et al. 1995, Thomsen et al. 1996
		Angora or Spanish goats and others	*Capra* spp.	Kouakou et al. 1992; Olkowski, Daar, and Olkowski 1991
		sheep	*Ovis aries*	Bell, Guerrero, and Granados 1996
gorse	*Ulex europaeus*	seed weevil	*Apion ulicis*	Markin, Yoshioka, and Brown 1995
Italian thistle milk thistle	*Carduus pycnocephalus* *Silybum marianum*	seedhead weevil	*Rhinocyllus conicus*	Goeden 1995a, 1995b
Klamath weed or St. Johnswort	*Hypericum perforatum*	root-boring beetle leaf beetle leaf beetle gall midge	*Agrilus hyperici* *Chrysolina quadrigemina* *Chrysolina hyperici* *Zeuxidiplosis giardi*	Campbell and McCaffrey 1991, Huffaker and Kennett 1959
leafy spurge	*Euphorbia esula*	stem and root borer	*Oberea erythrocephala*	Pemberton 1995b
Mediterranean sage	*Salvia aethiopis*	leaf and root chewing and mining weevil	*Phrydiuchus tau*	Andres, Coombs, and McCaffrey 1995
musk thistle or nodding thistle	*Carduus nutans*	seedhead weevil crown- and root-mining weevil	*Rhinocyllus conicus* *Trichosirocalus horridus*	Andres and Rees 1995
northern jointvetch	*Aeschynomene virginica*	Collego§	*Colletotrichum gloeosporioides* ssp. *aeschynomene*§	Smith 1986
perennial grasses and nutsedges	various spp.	Chinese weeder geese and others	*Picea glauca*	Wurtz 1995
puncturevine	*Tribulus terrestris*	seed-eating weevil crown- and stem-mining weevil	*Microlarinus lareynii* *Microlarinus lypriformis*	Andres and Goeden 1995
Scotch broom	*Cytisus scoparius*	seed-eating weevil twig-mining moth	*Apion fuscirostre* *Leucoptera spartifoliella*	Andres and Coombs 1995
spurge	*Euphorbia* spp.	leaf- and root-feeding flea beetles gall midge	*Aphthona cyparissiae* *Aphthona nigriscutis* *Spurgia esulae*	Pemberton 1995b
stranglervine or milkweed vine	*Morrenia odorata*	DeVine§	*Phytophthora palmivora*	
tansy ragwort	*Senecio jacobaea*	crown-, leaf-, and stem-feeding flea beetle shoot-feeding cinnabar moth	*Longitarsus jacobaeae* *Tyria jacobaeae*	McEvoy, Cox, and Coombs 1991; Pemberton and Turner 1990

TABLE 5-1. Selected biological control agents of weeds (*continued*).

| WEED | | BIOLOGICAL CONTROL AGENT | | |
Common name	*Scientific name*	*Common name*	*Scientific name*	*References*
tumbleweed or Russian thistle	*Salsola australis* (=*S. iberica*)	leaf-mining and chewing moth stem-boring moth	*Coleophora klimeschiella* *Coleophora parthenica*	Goeden and Pemberton 1995
yellow starthistle	*Centaurea solstitialis*	seed head weevil seed head gall fly	*Bangasternus orientalis* *Urophora sirunaseva*	Turner, Johnson, and McCaffrey 1995

Note: Most of the insects listed here were introduced for classical biological control of weeds. Natural populations of these insects are quite effective in some locations and may already be present locally. Distributors often collect these insects from the field and sell them, but there is little information on the weed control effectiveness of releasing these purchased insects. Contact your local county department of agriculture before purchasing or introducing any weed biological control organisms.

* Grass carp or white amur are prohibited in much of California. Consult your county department of agriculture for permitted uses.

§ Commercial availability is uncertain. Check labels for registration.

Sources: Cook et al. 1996, Hunter 1997, Julien 1992, Nechols et al. 1995.

crops that die when the harvested crop starts growing), selective planting (such as cover cropping only in otherwise empty rows between the harvested crop), and cover crop management (such as timed irrigation, mowing, cultivation and selective herbicide use).

Allelopathy is a form of antagonism or antibiosis in which plants release chemicals that retard growth of nearby plants. Certain cereal grains and aromatic shrubs, walnut, and many desert species (such as sagebrush) produce allelochemicals (Creamer et al. 1996). Mature plants often tolerate allelochemicals, but germination and seedling growth can be reduced. There are currently few specific recommendations for effectively using allelopathic plants to control weeds. Crop rotations and leaving certain crop residues in the field are among possible uses (Rice 1995).

■ **Pathogens.** Fungi and certain other microorganisms damage or kill certain weeds. Microorganisms called mycoherbicides (Hoagland 1990) are produced commercially and sprayed to control certain weed species infesting some crops (table 5-1). *Phytophthora palmivora* has been registered for stranglervine control in Florida citrus orchards. *Colletotrichum gloeosporioides* has been registered for control of

northern jointvetch infesting rice and soybeans in the central and southeastern United States; a single application as weeds emerge above the crop canopy can kill over 90 percent of the jointvetch by girdling the stem, although good control does not occur until about 4 weeks after application (Smith 1986). Despite their effectiveness, the market for these selective products is small and their commercial viability is uncertain.

Some introduced microorganisms have become naturalized. These include a rust fungus (*Puccinia chondrillina*) from Europe that now controls rush skeletonweed infesting rangelands and crops in certain areas of the western United States and Australia. Following introduction of *P. chondrillina* and two

This winter smother crop of yellow mustard minimizes weed growth in this vineyard. Cover crops also can help conserve beneficial insects by providing them with shelter and alternative food. Good management can minimize cover crop competition with the harvested crop. Cover crop selection is also important, in part because certain cover crops also can harbor pests, such as orange tortrix overwintering on mustard in apple orchards.

Well-managed livestock grazing can control many rangeland weeds, such as yellow starthistle.

<table>
<tr><td>⊢——⊣
LENGTH</td><td>The adult Klamathweed beetle is a shiny greenish blue to bronze color.
Klamathweed beetle and several other insects introduced to control toxic
Klamath weed save ranchers millions of dollars each year in otherwise
lost grazing land and poisoned livestock.</td></tr>
</table>

The adult Klamathweed beetle is a shiny greenish blue to bronze color. Klamathweed beetle and several other insects introduced to control toxic Klamath weed save ranchers millions of dollars each year in otherwise lost grazing land and poisoned livestock. *Hypericum calycinum*, an introduced ground cover closely related to *H. perforatum*, is popular in landscapes, where it can be defoliated by Klamathweed beetle. Klamathweed beetle control options in landscapes include periodic flooding or applying parasitic nematodes when beetles are pupating (usually in spring), removing litter that shelters adults during summer in hot areas, and spot application of insecticide to ornamental foliage to kill adults and larvae.

weed-feeding insects into northern California, density of rush skeletonweed was reduced from 60 to 90 % (Supkoff, Joley, and Marois 1988).

■ **Seed Bank Degradation.** Many weeds can produce thousands of seeds per plant each year. If seeds were not degraded by microorganisms (pathogens) or fed upon by animals (herbivores), weeds would be much more abundant. In comparison with deep plowing, shallow tilling or leaving seeds on the soil surface can reduce seed survival by 60 percent or more because seeds near the surface are more readily eaten by birds, insects, and rodents or decomposed by microorganisms.

■ **Vertebrates.** Birds, rodents, and many other small vertebrates rely heavily on seeds, fruit, roots, seedlings, or other plant parts as major components of their diet. Sheep may be the domesticated animal most widely used for weed

control, for example, to control weeds growing on rice levees and in seedling alfalfa (Bell, Guerrero, and Granados 1996). Goats can control poison oak (Kouakou et al. 1992) and can be crowded together to feed and clear areas of established vegetation, for example, to create firebreaks on steep hillsides that cannot be mowed or easily sprayed. Controlled grazing by cattle prior to the spine formation stage helps control yellow starthistle (Thomsen et al. 1996).

As many as 200,000 geese were used to weed cotton in California's San Joaquin Valley before the development of synthetic herbicides. These domesticated "weeder" geese have been found useful for eating weed seedlings in other established crops, including strawberries and nursery-grown conifers (Wurtz 1995).

Chinese grass carp and certain other fish are used in the biological control of aquatic weeds that interfere with water use or quality. Introducing fish is generally illegal in California because exotic fish might compete with and displace native species and game fish. Sterile fish, which can feed on weeds but are unable to reproduce (Chilton and Muoneke 1992), have been introduced to control aquatic weeds throughout the United States, including in Southern California irrigation canals. Check with your county department of agriculture to determine whether sterile fish introductions are permitted in your area.

■ **Invertebrates.** In many habitats, invertebrates (primarily insects, but also mites) are the most abundant group of plant-feeding animals. Biological control of weeds has emphasized the classical approach of introducing exotic species of insects that feed on introduced weeds. At least 268 species of insects (mostly beetles, flies, and moths) in 56 families have been introduced in attempts to control various weeds (Julien 1992, van Driesche and Bellows 1996). A notable success in the western United States is biological control of Klamath weed (=St. Johnswort, *Hypericum perforatum*), a rangeland weed that is toxic to livestock. Klamathweed beetle (*Chrysolina quadrigemina*) (fig. 5-1) and several other species introduced from Europe have largely eliminated Klamath weed from millions of acres in the western United States, Australia, Canada, Chile, New Zealand, and South Africa (Campbell and McCaffrey 1991, Huffaker and Kennett 1959). Other introduced insects in certain areas of the West have significantly reduced Mediterranean sage, puncturevine, Scotch thistle, tansy ragwort, and several other weeds infesting rangelands and uncultivated areas (Nechols et al. 1995). Alligatorweed in Gulf Coast states, musk or nodding thistle in the midwestern and eastern states, and waterlettuce in Florida are examples of other successful weed biological control projects (Julien 1992).

COMMERCIALLY AVAILABLE WEED BIOCONTROL AGENTS

Over two dozen species of insects (Hunter 1997), several fungi (mycoherbicides), and various vertebrates are commercially available for weed control in the United States. If monitoring indicates that the weed-eating species are absent locally, it may be beneficial to obtain and introduce them; check with your local agriculture agency to be sure introductions are permitted. In most areas of California, the county departments of agriculture distribute certain weed biological control agents. Learn about weed biology and conditions that favor natural enemies to determine when and how to best introduce them. Determine whether changes in cultural practices, such as irrigation, mowing, or avoiding mowing at certain times, may enhance natural enemy effectiveness.

INTEGRATING WEED BIOLOGICAL CONTROL AGENTS

Several mechanisms acting in combination can often control weeds more effectively than any single biological control agent. An outstanding example is prickly pear cacti, which were brought under biological control through a combination of competitive desirable plants and herbivorous invertebrates and vertebrates.

Exotic prickly pear cacti (*Opuntia* spp.) once infested about 50 million acres of Australian rangeland. Several

DIAMETER

Puncturevine's spiny seed capsules (center) injure people and grazing livestock and puncture tires. Two introduced weevils provide effective biological control of puncturevine in Hawaii and partial control in some areas of California (Andres and Goeden 1995). Adults are grayish to brown snout beetles. Their presence can be recognized by feeding scars (lighter patches on the stem and brownish areas on the green seed capsule) or by cutting open and inspecting for larvae, pupae, or frass in plant crowns or stems (for *Microlarinus lypriformis*) or seed heads (*M. lareynii*). An adult *M. lypriformis* emerged from the hole in this stem after feeding inside.

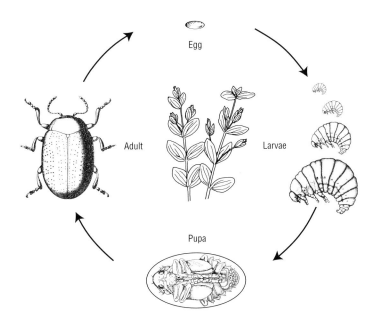

Egg

Adult

Larvae

Pupa

■ FIGURE 5-1. Klamath weed (*Hypericum perforatum*) is an excellent example of classical biological control of weeds in the western United States. The female Klamathweed beetle lays eggs singly or in small batches on young *Hypericum* leaves. The grayish larvae feed on foliage and develop through four instars before dropping or crawling to pupate within a cell they make just beneath the soil surface. Adults feed on foliage year-round, except during the hot, dry summer. Most damage occurs from larvae feeding during spring and early summer. Klamathweed beetle commonly has one generation a year, although multiple life stages may occur at one time and development varies with location and weather. Adapted from Holloway 1957 (adult, larva *from* Frank 1943; pupa *from* Chittenden 1921).

This weevil (*Eustenopus villosus*) and several other introduced insects reduce yellow starthistle reproduction and spread by feeding on seed heads or flowers (Turner, Johnson, and McCaffrey 1995). Researchers are seeking biological control of established starthistle infestations, which can currently be controlled by mowing to ground level during early flowering, intensive grazing during the bolting stages (May through June) before the spine formation stage, cultivation, burning, or herbicides followed by revegetation with fast-growing forage species to exclude starthistle regrowth (Lanini et al. 1995, Thomsen et al. 1996).

Whitish cochineal insects infesting a prickly pear cactus. Underneath their cottony covering, cochineal insects are a deep purple and are sometimes used to dye cloth, especially in Mexico (Donkin 1977). Removal of livestock, the resulting competition from the enhanced growth of other plant species, and this introduced insect now provide integrated biological control of prickly pear cacti on Santa Cruz Island, California (Goeden, Fleschner, and Ricker 1967).

insects were introduced to provide biological control, most notably the cactus moth (*Cactoblastis cactorum*). Moth larvae bore into plant tissue, stunting cactus growth. Prickly pear cacti in Australia are now under effective biological control because the boring larvae cause open wounds that become infected with cactus-killing microorganisms.

On Santa Cruz Island in California, excessive grazing denuded the island's grasslands, allowing native prickly pear cacti to become overabundant. A cochineal insect was introduced to stunt cactus growth. Feral sheep were removed and cattle grazing was better managed and eventually stopped. The introduced cochineal insect and competition from the enhanced growth of other plant species now act in combination to provide biological control of prickly pear cacti on the Island (Goeden, Fleschner, and Ricker 1967).

Natural enemies of prickly pear cacti have not been deliberately introduced into the continental United States because some native prickly pear species are ecologically valued and endangered with extinction. Cactus moth, which was accidentally introduced into Florida in the early 1990s (Pemberton 1995a), is expected to spread and eventually reduce prickly pear cacti populations in the West.

A *Microlarinus lypriformis* larva (with brown head) and pupa exposed in their tunnels. Female *M. lypriformis* lay eggs in puncturevine terminals and on the underside of stems and leaves. The hatching larvae stunt and kill plants as they bore and feed inside stems and roots. *M. lareynii* adults chew holes and lay eggs in developing puncturevine seeds, where their whitish larvae feed and pupate. Both species have about 2 to 4 generations each year, and in the absence of prolonged freezing weather provide good control in uncultivated areas where insecticides are not applied. Puncturevine biological control is poor in cultivated croplands and where insecticides are sprayed.

CHAPTER SIX

NATURAL ENEMIES OF ARTHROPODS

Arthropods (such as insects and mites) and other invertebrate pests (such as nematodes and snails) damage or kill plants primarily by feeding on them. Invertebrates sometimes interact with other types of pests to cause even greater damage, such as by vectoring viruses or other plant pathogens. Because most biological pest control work has focused on arthropods (invertebrates such as insects and mites that have jointed appendages and a hard outer skin), this book emphasizes natural enemies of arthropods. Biological control of arthropods relies on predators, parasites, or pathogens manipulated through conservation and enhancement, augmentation, and classical biological control.

CLASSICAL BIOLOGICAL CONTROL OF ARTHROPODS

Over 400 species of insects worldwide reportedly have been completely or partially controlled through introductions of exotic natural enemies (Greathead and Greathead 1992, Laing and Hamai 1976, Luck 1981). Pest Homoptera (such as scale insects and whiteflies) have most frequently been targets of successful introductions of predators and parasites (Hall and Ehler 1979, Stiling 1990). However, examples of successful classical biological control can be found for at least a few species in most major insect groups. Introductions have also been somewhat successful in controlling certain mites and the brown garden snail.

CONSERVATION AND ENHANCEMENT OF ARTHROPODS

Conserve resident beneficials whenever you can by choosing cultural, physical, and selective chemical controls that do not interfere with or kill natural enemies. Most pests are attacked by a complex of natural enemies (table 6-1), and their conservation

This adult assassin bug (*Zelus renardii*) is a common predator. It impales insects with its piercing mouthparts and injects a paralyzing venom.

41

NATURAL ENEMIES	PESTS								
	thrips	spider mites	aphids	whiteflies	caterpillar eggs	plant bugs	leafhoppers	bollworm	beet armyworm
Predators									
green lacewing (*Chrysoperla carnea*)	■	■	■	■	■			■	■
minute pirate bug (*Orius tristicolor*)	■	■	■	■	■			■	■
bigeyed bug (*Geocoris pallens*)		■			■	■	■	■	■
damsel bug (*Nabis americoferus*)			■			■	■	■	■
Parasitic wasps									
Cotesia marginiventris								■	■
Hyposoter exiguae								■	■
Chelonus texanus								■	■
Trichogramma semifumatum								■	■

Note: These natural enemies of the cabbage looper are often more abundant in cotton when looper is present; each ■ indicates other pests attacked by these predators and parasites. Moderate looper populations may be more beneficial than harmful, as loopers support populations of the natural enemies that help to control these other pests. Cabbage looper is a secondary pest of cotton that causes significant damage only when its natural enemies are disrupted, such as by broad-spectrum pesticides applied against other pests. If treatment of loopers is necessary, a selective material such as *Bacillus thuringiensis* is a good choice for minimizing disruption of natural enemies. Additional important species of natural enemies that in cotton primarily or only attack the cabbage looper are several parasitic wasps (*Copidosoma truncatellum, Microplitis brassicae, Patrocloides montanus*), a parasitic fly (*Voria ruralis*), and a nuclear polyhedrosis virus. Consult the index to find where these species are discussed.

Sources: Anonymous 1996, Ehler 1977, Ehler and van den Bosch 1974.

is the primary way most growers and gardeners can successfully use biological control. Pesticide management, ant control, and habitat manipulation are the key conservation strategies summarized here.

■ **Pesticide Management.** Pesticides can severely disrupt biological control. Natural enemies are often more susceptible to pesticides than pests are due to many ecological and physiological factors (Hull and Beers 1985). In comparison with most pests, natural enemies are more active searchers, which results in greater contact with treated surfaces. Migratory natural enemies may not reside in treated growing areas year-round, so they are less subject to the constant pesticide selection pressures that promote pesticide resistance in permanently residing pest populations. Because natural enemies require pests as hosts, few will be present after pest numbers are reduced by spraying. Pests thereby get a head start in reproducing, and natural enemy populations will lag behind until enough pests develop to attract natural enemies and support their reproduction. Also, plants produce toxic secondary chemicals in their leaves and wood to ward off plant-feeding herbivores. Pests have evolved physiological systems to detoxify and protect themselves from plant chemicals; these same chemical defenses help pests develop pesticide resistance. Natural enemies, however, often lack these chemical-defenses because their food (the pests) often lack chemicals that are poisonous to other invertebrates.

In addition to immediately killing natural enemies present at the time of spraying (contact toxicity), many pesticides leave persistent residues that kill predators or parasites that migrate in long after spraying (residual toxicity) (Theiling and Croft 1988). Even if beneficial organisms survive an application, low levels of pesticide residues can have adverse sublethal effects on natural enemy longevity, fecundity, and ability to locate and kill pests (Rosenheim and Hoy 1988; Umoru, Powell, and Clark 1996).

Avoid applying broad-spectrum, persistent pesticides where natural enemies are present. Tables 1-2 and 6-2 and figure 6-1 summarize the relative toxicity of different types of pesticides to natural enemies. When pesticides are used, apply them in a selective manner (such as spot applications), time applications to minimize impacts on natural enemies, and choose non-persistent materials or selective materials (those that are more specific in the types of pests they kill).

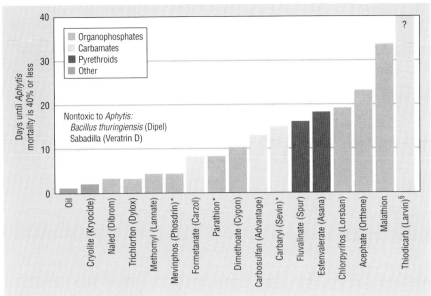

Argentine ants are tending this black scale colony, preventing the scales from being attacked by natural enemies that can control black scale. Other species of ants are sometimes beneficial predators of pests. Observing ants' behavior may tell you whether the ants are a pest.

■ **Ant Control.** Ants are beneficial as consumers of weed seeds, predators of many insect pests, soil builders, and nutrient cyclers. Some ants are direct pests of crops, feeding on nuts or fruit. The common Argentine ant and certain other species are pests because they feed on honeydew produced by aphids, soft scales, whiteflies, mealybugs, and some other Homoptera; these ants protect the Homoptera from predators and parasites that might otherwise control these pests (Bartlett 1961), and ants may move these honeydew-producing insects from plant to plant. For example, merely preventing ants from foraging on mealybug-infested grapevines will reduce mealybug populations and their damage (Phillips and Sherk 1991). Certain ants can also disrupt the biological control of non-honeydew-producing pests, such as mites (Haney, Luck, and Moreno 1987) and armored scales (Annecke 1959), if these pests occur on the same plants as honeydew-producing species. Where natural enemies are present, if ants are controlled, populations of many pests will be reduced gradually as

■ FIGURE 6-1. The number of days after application of various insecticides until 40 percent or less of exposed red scale parasites (*Aphytis melinus*) were killed (Bellows et al. 1985, 1993; Campbell 1975). Some pesticides (such as many carbamates, organophosphates, and pyrethroids) have residual toxicity that can kill natural enemies long after application. Even if natural enemies are not killed, pesticide residues can have sublethal effects that decrease natural enemies' effectiveness (see Rosenheim and Hoy 1988). Reported toxicities are only a general guide; actual toxicity varies depending on environmental conditions, application rate, exposure, and the species of natural enemy, as well as the particular pesticide.

*Mortality in the study of mevinphos, parathion, and some formulations of carbaryl may have persisted longer than reported here.

§ Thiodicarb continued to kill more than about 80 percent of parasites after 36 days, when testing was stopped.

Argentine ants attracted to poison bait in a container plant. Worker ants, such as those emerging from the hole in this stake, are about ¹⁄₁₀ inch (2.5 mm) long and are uniformly dark brown. They are often observed following trails.

TABLE 6-2. Toxicity to natural enemies of selected insecticides and acaricides.

PESTICIDE CHEMICAL NAME (TRADE NAMES)	CLASS	RANGE OF ACTIVITY (AFFECTED GROUPS)	IMMEDIATE IMPACT ON NATURAL ENEMIES	DURATION OF IMPACT ON NATURAL ENEMIES
abamectin (Avid, Zephyr)	M	moderate (mites, leafminers)	high to predatory mites, low for many insects	long to predatory mites and affected insects
acephate (Orthene)	OP	broad (insects and mites)	high	intermediate
aldicarb (Temik)	C	broad (insects and mites)	moderate (at planting) low (sidedress)	moderate (at planting) low (sidedress)
azinophosmethyl (Guthion)	OP	broad (insects and mites)	high	long
azadirachtin (Margosan-O, Neemix)	B, IGR	broad (insects and mites)	moderate	short
Bacillus thuringiensis ssp. *kurstaki* (Dipel, Thuricide)	M	narrow (caterpillars)	none	none
Bacillus thuringiensis ssp. *israelensis* (Bactimos, Gnatrol)	M	narrow (larvae of fungus gnats, mosquitoes, some other flies)	none	none
Bacillus thuringiensis ssp. *san diego* or *tenebrionis* (M-Trak, Novodor)	M	narrow (leaf beetles)	none	none
bendiocarb (Dycarb, Ficam)	C	broad (insects and mites)	high	high
bifenthrin (Capture, Talstar)	P	broad (insects and mites)	high	long
carbaryl (Sevin, XLR Plus)	C	broad (insects and mites)	high	long
carbofuran (Furadan)	C	broad (insects and mites)	high	intermediate
chlorpyrifos (Dursban, Lorsban)	OP	broad (insects and mites)	high	intermediate
copper bands	CON	narrow (snails and slugs)	none	none
cryolite (Kryocide)	I	narrow (foliage chewing insects)	low to none	low to none
cyfluthrin (Baythroid, Tempo)	P	broad (insects and mites)	high	intermediate
cypermethrin (Ammo)	P	broad (insects and mites)	high	intermediate
diazinon	OP	broad (insects and mites)	high	intermediate to high
dichlorvos (DDVP, Vapona)	OP	broad (insects and mites)	high	intermediate to high
dicofol (Kelthane)	CH	narrow (pest mites and mites)	high to beneficial mites	long to beneficial mites
dimethoate (Cygon)	OP	broad (insects and mites)	high	long
fenbutatin-oxide (Vendex)	OT	narrow (pest mites)	low	short
fenpropathrin (Tame)	P	broad (insects and mites)	high	intermediate
fenitrothion (Danitol)	OP	broad (insects and mites)	high	intermediate
formetanate hydrochloride (Carzol)	C	broad (insects and mites)	high	long, unless washed off
Gossyplure	PH	narrow (pink bollworm)	none	none
insecticidal soap (M-Pede, Ringers, Safer's)	CON	broad (insects and mites)	moderate	short to none
imidacloprid (Admire, Marathon, Merit, Provado)	N	narrow (sucking insects)	low	short
kinoprene (Enstar)	IGR	intermediate (immature stage insects)	moderate to low	low
lime, hydrated	I	narrow (leafhoppers)	low	moderate
malathion	OP	broad (insects and mites)	high	intermediate
metaldehyde (Deadline)	A	narrow (pest snails and slugs and beneficial snails)	none, except to decollate snail	none, except to decollate snail

PESTICIDE CHEMICAL NAME (TRADE NAMES)	CLASS	RANGE OF ACTIVITY (AFFECTED GROUPS)	IMMEDIATE IMPACT ON NATURAL ENEMIES	DURATION OF IMPACT ON NATURAL ENEMIES
methamidophos (Monitor)	OP	broad (insects and mites)	moderate	intermediate
methidathion (Supracide)	OP	broad (insects and mites)	high	intermediate to long
methomyl (Lannate)	C	broad (insects and mites)	high	intermediate
naled (Dibrom)	OP	broad (insects and mites)	high	intermediate
oil (SunSpray)	CON	broad (exposed insects and mites)	moderate	short to none
oxamyl (Vydate)	C	broad (insects and mites)	high	intermediate
oxydemeton-methyl (Metasystox-R)	OP	narrow (sucking insects and mites)	high to beneficial mites low to insects	intermediate to mites, low to none to insects
permethrin (Ambush, Pounce, Pramex)	P	broad (insects and mites)	high	long
profenofos (Curacron)	OP	broad (insects)	moderate	short
propargite (Comite, Omite)	OS	narrow (pest mites)	low to none	short to none
pyrethrum + piperonyl butoxide (Pyrenone)	B	broad (insects)	high	short
resmethrin	P	broad (insects and mites)	high	intermediate
rotenone	B	narrow (aphids and some soft scales)	moderate to none	short to none
ryania	B	narrow (citrus thrips)	low to none	short to none
sabadilla (Veratrin D)	B	narrow (citrus thrips)	low to none	short to none
sticky materials (Tanglefoot, Stickem)	CON	narrow (trunk climbing insects and snails)	moderate to none	long to none
sulprofos (Bolstar)	OP	broad (insects)	high to beneficial insects	intermediate to short
sulfotep (Plantfume 103)	OP	broad (insects and mites)	high	intermediate to high
sulfur	I	narrow (mites and citrus thrips)	moderate to beneficial mites	intermediate to mites

KEY			
A	acetaldehyde	M	microbial
B	botanical	N	chloronicotinyl nitroguanidine
C	carbamate		
CH	chlorinated hydrocarbon	OP	organophosphate
		OS	organosulfur
CON	contact including smothering and barrier effect	OT	organotin
		P	pyrethroid
		PH	pheromone or mating disruptant
I	inorganic		
IGR	insect growth regulator		

Note: The immediate impact of pesticides on natural enemies is the killing of natural enemies resulting from spraying the pest or its habitat (contact toxicity). Duration of the impact on natural enemies refers to persistent residues that kill natural enemies that migrate in and contact previously treated areas (residual toxicity). Stated toxicities should be used only as a general guide. The actual toxicity of specific chemicals depends on environmental conditions, application rate and exposure, and the species of natural enemy. The most widely tested natural enemy species were *Phytoseiulus persimilis* predatory mites; various predatory lady beetles and green lacewings; and *Aphytis, Encarsia,* and *Trichogramma* spp. parasitic wasps.

Sources: Croft 1990, Grafton-Cardwell et al. 1996, Hassan et al. 1994, Jepson 1989.

natural enemies become more abundant (Vander Meer, Jaffee, and Cedeno 1990; Way and Khoo 1992).

Ant control methods include cultivation, barriers, and insecticide baits or insecticide sprays around the base of plants. Research in orchards and vineyards demonstrates that the Argentine ant can be controlled with farnesol, a behavior-modifying chemical (pheromone) or repellent (Shorey et al. 1996, Sisk et al. 1996); watch for new information on and the commercial availability of this product.

Deny ants access to plants by pruning branches that provide a bridge between buildings, other plants, or the ground and by applying sticky material (such as Tanglefoot or Stickem) to encircle trunks (Phillips, Bekey, and Goodall 1987). Sticky material, slippery Teflonlike sprays or wraps, or moats of water or oil will exclude ants from container plants or growing benches. Spraying the soil surface or drenching soil around nests with certain insecticides also reduces ant populations.

Broad-spectrum or volatile pesticides sprayed to control ants can harm natural enemies.

Place enclosed pesticide baits, such as ant stakes, near nests, the base of plants, or along ant trails (Baker, Key, and Gaston 1985; Knight and Rust 1991). Pesticide baits act slowly but are more selective than sprays and can be more effective because worker ants carry the poisoned bait back to their underground nests, killing the reproductive queens and causing the entire colony to die. In California, treat ants in the late winter or early spring when ant populations are low. Periodic moistening and stirring some baits may improve their attractiveness to ants.

■ **Habitat Manipulation.** Good management of plants and growing areas can enhance natural enemy effectiveness (Flint and Roberts 1988). Border plants or vegetation interplanted with crops can provide natural enemies with many resources (such as nectar, pollen, and shelter) and serve as a source of natural enemies that migrate in and help control crop pests (Corbett and Plant 1993). For example, alfalfa can be managed to improve crop diversity and enhance biological control. Alfalfa is a tremendous reservoir of natural enemies that migrate out and help control pests in surrounding crops. However, harvesting the entire crop of alfalfa at one time drives pests such as lygus bugs (which do not cause economic damage to alfalfa) into nearby cotton or bean fields, where lygus can cause serious damage. By leaving a border of alfalfa unharvested along each row, lygus bugs remain in the alfalfa, and the crop continues to produce beneficials (fig. 6-2) (Stern 1969, Summers 1976).

Adult parasites and the adult stage of many insects with predaceous larvae, such as green lacewings and syrphid flies, feed only on pollen and nectar. Even if pests are abundant for the predaceous and parasitic stages, many beneficials will do poorly unless flowering or nectar-producing plants are available to adult natural enemies (fig. 6-3).

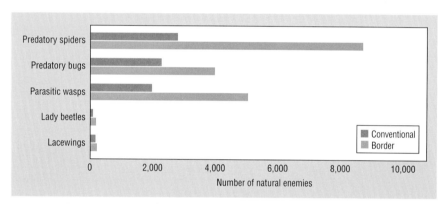

■ FIGURE 6-2. Comparison of the relative abundance of naturally enemies in a border versus a conventional cut alfalfa field over a 4-month period from May through September. Data from Summers 1976.

	Moisture*	Jan.	Feb.	Mar.	Apr.	May	June	July	Aug.	Sept.	Oct.	Nov.	Dec.
Willow species	W												
Ceanothus spp.	D												
Redbud	D-I												
Mule fat	I-W												
Yarrow species	D-I												
Coffeeberry	D-I												
Hollyleaf cherry	I												
Soapbark tree	I												
Buckwheat species	D												
Elderberry species	I-W												
Toyon	D												
Creeping boobyalla	I												
Bottletree	I												
Narrowleaf milkweed	D-I												
Coyote brush	D-I												

*Moisture requirements:
dry (D)   dry to intermediate (D-I)   intermediate (I)   intermediate to wet (I-W)   wet (W)

■ FIGURE 6-3. Darkened cells show the flowering periods of some perennial insectary plants that, when used in the right combination, can provide nectar and pollen for natural enemies sequentially throughout the year in California (Anonymous 1995b, Bugg and Anderson n.d.). Avoid insectary plants that are reservoirs of viruses or alternate hosts of other pests that can damage your crops; consult publications such as *Pests of Landscape Trees and Shrubs* (Dreistadt 1994) and *Insects that Feed on Trees and Shrubs* (Johnson and Lyon 1988) to identify alternate hosts of your pests. See Steffan and Whitaker (1996) for insectary plants suggested for the midwestern United States.

Sticky material around trunks can exclude flightless pests such as ants and weevils. Rather than applying sticky material directly on bark, it is often a good idea to wrap trunks with some flexible material then apply the sticky barrier onto the trunk wrap.

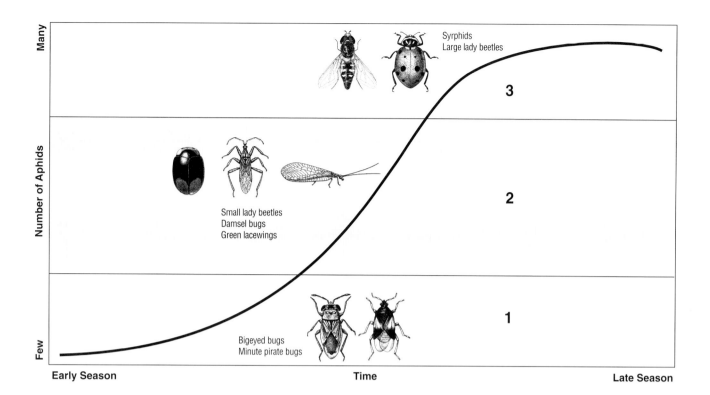

■ FIGURE 6-4. The most important species of predators vary depending on pest density and the length of time that pests have been present, as shown here for green peach aphid infesting sugarbeets. Pest populations often develop over time through three phases of relative abundance: (1) beginning growth from low density, (2) rapid growth, and (3) peak density and decline. The predaceous bugs important in sugarbeets during phase 1 (low aphid density) are general feeders that prey on many different species of pests. The predators most important during phase 3 commonly are host specific; the convergent lady beetle and syrphids feed primarily on aphids. Parasite species and abundance can also vary, while pathogens become more important at higher pest densities. Because aphids vector plant viruses, growers may not tolerate many aphids in sugarbeets, but this pattern of pest and natural enemy development applies to many situations. *Sources:* Tamaki 1981, Tamaki and Long 1978 (bugs, lacewing, and large lady beetle by Celeste Green *from* Smith and Hagen 1956; small lady beetle *from* Gordon 1985, reprinted with permission from the American Museum of Natural History; syrphid *from* Cole 1969, by C. S. Papp, reprinted with permission from University of California Press).

Alfalfa intercropped in cotton fields attracts lygus bugs, which are harmless to alfalfa hay and would otherwise feed on and damage the cotton. Strip-cutting alfalfa (harvesting only part of the crop during each cutting) improves its value as a refuge for natural enemies and discourages lygus bugs from migrating to cotton (Stern 1969, Godfrey and Leigh 1994).

Plant diversity as shown in this garden can enhance biological control by providing natural enemies with shelter, nectar, pollen, and alternate prey.

Growing a variety of diverse plant species (Altieri 1994) or certain insectary plants can provide natural enemies with food and shelter and increase the fecundity, longevity, and population density of beneficials (Idris and Grafius 1995, 1996; Steffan and Whitaker 1996). Cover crops, secondary plantings grown not for harvest but to enhance production of the primary crop, are widely used in orchards and vineyards (Bugg and Waddington 1994, Bugg et al. 1996, Smith et al. 1996). If properly managed, cover crops can provide many benefits. See *Orchard Floor Mulching Trials* (Lanini, Shribbs, and Elmore 1988) and *Covercrops for California Agriculture* (Miller et al. 1989).

Beneficial organisms require vegetation to hide from their own natural enemies, overwinter in, and be protected from adverse weather or environmental conditions, such as the heat of summer afternoons. Diverse vegetation generally harbors low populations of many different arthropods that may serve as alternate hosts for natural enemies (Bugg, Ehler, and Wilson 1987; Bugg and Wilson 1989, Starý 1991) and help maintain natural enemies so that they remain in the local habitat and are present to help prevent and control pest outbreaks (fig. 6-4, table 6-1). Commercially available attractants and artificial foods may attract and feed certain beneficials (see the discussion of lacewings in chapter 8). There is little information on the practical or economic effectiveness of applying attractants or using cover crops or perennial borders to improve biological control.

Dust interferes with some predators' and parasites' ability to locate prey and feed on or parasitize hosts. Excess dust can cause outbreaks of pests such as spider mites. Reduce dust by growing windbreaks, planting bare soil, sprinkling dirt surfaces with water before they experience heavy traffic, driving slowly on dirt roads around fields, installing plastic sheets or other barriers between dust sources and crops, and minimizing cultivation and other disturbances of dry soil near crops.

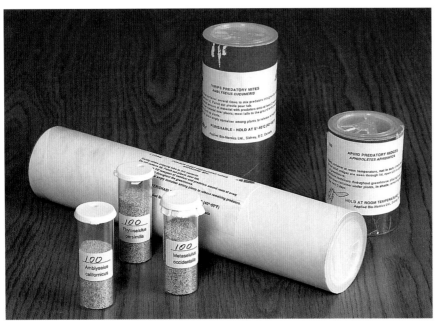

Many natural enemies are available for purchase and are shipped through mail services in containers such as these. If resident natural enemies are insufficient, releases can augment biological control in certain situations.

This card contains parasitized whitefly pupae. It can be hung on plants to introduce tiny parasitic wasps in greenhouses or interior plantscapes, such as indoor malls and large office buildings.

AUGMENTATION OF ARTHROPODS

There is a great deal of interest in purchasing and releasing natural enemies for augmentation to increase biological control, and many species are available for purchase and delivery through mail-order services (table 6-3). However, in most situations, pests cannot be effectively managed by purchasing and releasing natural enemies.

For augmentation to be effective in most cases, releases must be combined with other compatible IPM practices. Releases are most likely to be effective in

Selected commercially available predators, parasites, and pathogens of invertebrates.

PEST	NATURAL ENEMY		
	Common name	*Scientific name*	*Page*
aphids	lady beetle	*Hippodamia convergens,* possibly others	82
	lacewings	*Chrysoperla* spp.	100
	minute pirate bugs	*Orius* spp., possibly others	91
	parasitic wasps	*Aphidius* spp. and others	66
	predaceous midge	*Aphidoletes aphidimyza*	95
broad mites	predaceous mites	*Neoseiulus* spp.	106
brown garden snail	predatory snail*	*Ruminia decollata**	115
caterpillars	egg parasites	*Trichogramma* spp.	58
	entomopathogenic nematodes	*Heterorhabditis bacteriophora, Steinernema carpocapsae*	119
	larval parasites	several host-specific species	61, 73
	microbial insecticide	*Bacillus thuringiensis* ssp. *kurstaki*	118
	nuclear polyhedrosis viruses	various species, *see* table 9-1	118
fungus gnats	predaceous mite	*Hypoaspis miles*	108
	microbial insecticide	*Bacillus thuringiensis* ssp. *israelensis*	118
	entomopathogenic nematodes	*Steinernema carpocapsae, S. feltiae*	119
leaf beetles	microbial insecticide	*Bacillus thuringiensis* ssp. *san diego* or *tenebrionis*	118
manure flies	parasitic flies and wasps	*Muscidifurax, Spalangia,* and other species	
mealybugs	lacewings	*Chrysoperla* spp.	100
	mealybug destroyer	*Cryptolaemus montrouzieri*	87
	citrus mealybug parasites	*Leptomastix dactylopii,* possibly others	70
mosquitoes	predatory mosquito fish	*Gambusia affinis*	
	microbial insecticides	*Bacillus sphaericus, Bacillus thuringiensis* ssp. *israelensis, Lagenidium giganteum*	118
scale insects	predaceous lady beetle	*Rhyzobius* or *Lindorus lophanthae,* possibly others	85
	red scale parasite	*Aphytis melinus*	65
	soft scale parasites	*Metaphycus helvolus, Microterys flavus*	65
serpentine leafminer	parasitic wasps	*Dacnusa, Diglyphus* spp.	72
spider mites	lacewings	*Chrysoperla* spp.	100
	predaceous mites	*Amblyseius, Galendromus, Metaseiulus, Phytoseiulus* spp.	106
thrips	greenhouse thrips parasite	*Thripobius semiluteus*	72
	lacewings	*Chrysoperla* spp.	100
	minute pirate bug	*Orius tristicolor*	91
	predaceous mites	*Amblyseius, Euseius* spp.	106
whiteflies	lacewings	*Chrysoperla* spp.	100
	parasitic wasps	*Encarsia, Eretmocerus* spp. and others	68
	predaceous lady beetle	*Delphastus pusillus*	87

Note: Release of natural enemies has not been consistently successful in controlling pests in most situations. Only the better-studied species are listed here. See table 9-1 for commercially available insect pathogens. See table 3-1 for commercially available natural enemies of plant pathogens. See the suppliers at the end of the book for additional species and sources of natural enemies.

*Legal in California only for release in certain southern counties. For more information on snails as predators, see chapter 8.

Sources: Cranshaw, Sclar, and Cooper 1996; Hunter 1997.

situations similar to those where researchers or pest managers have previously demonstrated success. Typically, this includes situations where certain levels of pests and damage can be tolerated. Augmentation may be more effective for perennial plants of relatively high value. Desperate situations where pests or damage are already abundant cannot usually be managed with augmentation.

Inundation (releasing large numbers to rapidly obtain control) and inocula-tion (releasing relatively small numbers that reproduce and eventually provide control through their progeny) are the two main release tactics. Releasing the mealybug destroyer lady beetle, as discussed in chapter 8, is an example of inoculative release. The mealybug destroyer overwinters poorly outdoors in California and often needs to be reintroduced to target areas in the spring. Periodically releasing *Trichogramma* spp. wasps to kill caterpillar eggs, such as those of codling moth or tomato fruitworm, or releasing lacewings or convergent lady beetles for aphid control, are examples of inundative biological control. However, success from these releases has been mixed. Although releases often rely on purchased arthropods, it may be feasible for some pest managers to rear certain natural enemies on-site on nurse plants.

■ **Releasing Natural Enemies Effectively.** Most practical experience with augmentation of natural enemies of arthropods comes from vegetable crops grown in greenhouses (Hussey and Scopes 1985, Malais and Ravensberg 1992) and work in a few field or orchard crops such as citrus (Haney et al. 1992). Many programs have not been carefully studied to document their cost and effectiveness. Although much information has been gathered in small-scale research projects, there are relatively few research-based recommendations for effectively releasing commercially available natural enemies of arthropods on a large scale.

Take steps to increase the likelihood that natural enemy releases will be effective. Accurately identify the pest and its life stages. Learn about the biology of the pest and its natural enemies. Anticipate pest problems and plan releases ahead of time; begin making releases before pests are too abundant or intolerable damage is imminent.

Release the appropriate natural enemy species when the pest is in its vulnerable life stages (most parasitic insects attack only certain stages of their hosts). Many parasites lay their eggs in one life stage of the host, but the parasite does not kill its host and emerge until the host has developed to another stage of the host. For example, the exit holes in mature female scales are often caused by progeny of para-

sites that laid their eggs in the host when it was immature. The pest life stage that can be effectively controlled with natural enemies may be different from the pest stage that damages plants. For example, most *Trichogramma* species kill only insect eggs, especially those of moths and butterflies; they are not effective against caterpillars. *Trichogramma* must be released when moths are laying eggs, before caterpillars and damage become abundant.

To reduce toxicity to natural enemies, either avoid applying broad-spectrum or persistent pesticides or use them as spot sprays. Pyrethroids, carbamates, and organophosphates are especially toxic to natural enemies (fig. 6-1, table 6-2). Test for possible toxicity before releasing natural enemies on plants sprayed with materials of unknown toxicity. For example, before releasing the parasitic wasp *Aphytis melinus* against California red scale, confine about 1,000 *Aphytis* in a gallon jar with 10 to 12 one-year-old (green) citrus twigs freshly clipped from your trees. If more than about 20 to 35 percent of parasites die within 24 hours, pesticide residues are too high for

release. In some situations, pesticide residues toxic to certain natural enemies may persist for months. If pesticide application is needed prior to releases, apply low-persistence materials and delay releases until testing shows that residues have degraded. To assess and compare parasite vigor, confine some parasites in a control jar filled with untreated leaves from trees that have never been sprayed.

When using augmentation for pest control, be prepared to tolerate some pests and minor damage since pests must be present to provide food for natural enemies. Remember that natural enemies are living organisms that require water, food, and shelter. Natural enemies may be adversely affected by extreme conditions such as hot temperatures. Keep them in a cool place and release them early or late in the day if temperatures are hot.

Many species of natural enemies stop reproducing under short day length or prolonged cool weather (van Houten et al. 1995). In greenhouses and interior plantscapes, supplemental light or heat may be necessary for some beneficials to be effective year-round.

Predatory mites are being released into a strawberry field using this tractor-mounted applicator. This equipment was developed by the Department of Agricultural and Biological Engineering at the University of California, Davis (Gardner and Giles 1996a, 1996b) and was funded in part by the UC Statewide IPM Project. This equipment is being commercialized by a private company. Photo by Ken Giles.

Effectively releasing natural enemies requires knowledge, practice, and imagination. The most effective control may be obtained when more than one species of parasite and predator are released in combination with other compatible control methods. Releases often fail because information or experience is inadequate, the wrong species is released, pesticides are applied, the beneficial organisms were shipped or handled incorrectly prior to release, or the timing was incorrect. For example, without adequate prerelease planning and preparation, by the time purchased natural enemies are delivered, it may be too late to effectively release them.

Obtain beneficial arthropods from a quality supplier. Find out when and how they will be shipped and know what day they are expected to arrive. Upon arrival, examine the natural enemies carefully to learn to distinguish them from other species. Learn how to store them until they can be released and release them as soon as conditions are suitable.

The quality of commercially available natural enemies is not regulated and may sometimes be poor due to production practices, inadequate packaging, or unsuitable conditions during shipment. Evaluate the quantity and quality of each shipment of natural enemies (Cranshaw, Sclar, and Cooper 1996; O'Neil 1997). Take samples to estimate numbers or emergence. For example, if parasites arrive inside hosts, count adult parasite exit holes immediately after receipt. Recount and compare the number of emergence holes about 10 days after releasing the parasites, or place a card or leaf containing parasitized hosts in each of several clear containers and inspect these and estimate emergence at various times after receipt. Besides emergence or immediate viability upon arrival, sex ratio, longevity, and size of individual insects are among those factors that can influence the effectiveness of parasites and predators.

Available natural enemies may not always be able to keep pest populations below acceptable damage thresholds. Many natural enemy species are com-

Natural enemy introductions must generally be combined with other effective pest-control methods. Special insect screens have been installed on this greenhouse to contain beneficials and prevent inward pest migration. Although screens with large holes can exclude larger pests such as moths, special screens and modifications in the ventilation system are required to exclude smaller species. Maximum hole sizes for excluding *Liriomyza* leafminers and western flower thrips range from about 640 microns to 192 microns respectively (Bethke and Paine 1991; Bethke, Redak, and Paine 1994).

mercially available largely because they are the easiest and most economical species to produce and sell. Unfortunately, they are not always the most effective species for an individual pest. In some cases, the value of the crop or host plants and availability of alternatives may not justify the cost and effort of releasing natural enemies. Certain beneficial predators, parasites, and pathogens that can be purchased and released to control common invertebrate pests are listed in tables 6-3, 7-3, and 9-1.

Introduction methods vary depending on the pest, crop, and growing conditions and on the species, life stage, and packaging of the natural enemies. Specific strategies for augmentative releases are revised as new information becomes available. Consult Cooperative Extension advisors, suppliers, and publications cited in chapters 7, 8, and 9 for specific release recommendations.

■ **Nurse Plants.** Banker plants or nurse plants are an economical way for sophisticated users to produce natural

enemies for their own use in certain situations. Nurse plants are plants infested with pests that are used on-site to rear natural enemies for release into crops. Natural enemies produced on-site are more readily available when needed and users can rear species or strains that may not be commercially available. By rearing their own beneficials, users become more familiar with natural enemy identification and biology and may be able to produce natural enemies that are better adapted for local conditions because the initial beneficial population can be collected in that area.

The major impediment is that on-site production usually requires growing susceptible plants, rearing the target pest or suitable alternative host, and producing sufficient numbers of good quality beneficials. Problems in producing any one of these three (plant, pest, or beneficial) can frustrate the entire project. Nurse plants are most feasible where relatively simple and proven procedures can be used to produce the natural enemies. For example, the

A portion of this greenhouse has been screened so that the grower can start small by releasing natural enemies on only a portion of the crop. A scout is exiting the biological control area through insect screen that has been overlapped to provide a doorway that excludes pests.

mealybug destroyer lady beetle or mealybug parasites can be produced on mealybug-infested *Coleus* or in wide-mouth jars or buckets containing sprouted potatoes infested with mealybugs (Fisher 1963). Mealybug parasites and predators can fly but mealybugs cannot, allowing self-distribution of the beneficials. A band of Vaseline, sticky material, or other barrier inside near the top of the container prevents mealybugs from escaping while allowing the parasitic wasps and adult lady beetles to fly away to nearby crops.

In certain situations, natural enemies can be introduced directly from nurse plants without potentially introducing pests. For example, the whitefly parasite *Encarsia formosa* can be reared within an enclosure screened with Lumite 2:1 twill weave polyethylene fiber, 0.116 mm² hole size. This screen prevents whiteflies from escaping and infesting plants, but allows the smaller parasites to pass through and parasitize whiteflies on crops (Bethke, Redak, and Paine 1994).

Aphidoletes predators and certain parasites attack many different species of aphids. These natural enemies can be introduced by rearing them on an alternative plant species that is scattered throughout the crop and heavily infested with species of aphids that do not attack the crop (Starý 1993). Rearing certain predatory mites on mite-infested bean plants (Cushing and Whalon 1986) or on bean pollen without host mites is another system used to rear natural enemies.

When crop plants heavily infested with a target pest are encountered, instead of treating or disposing of plants, consider caging or covering infested plants and (if in containers) moving them from the general growing area to an isolated or enclosed location. Observe the plants to determine whether they have been naturally colonized by locally occurring beneficial species or strains that can be reared or purchase or collect some of the appropriate natural enemies and introduce and confine them on the infested plants. Once the natural enemies reproduce and become abundant, collect and release them into the crop or scatter the nurse plants throughout the growing area and uncage them so that adult natural enemies can disperse into the crop.

Grower production of *Phytoseiulus persimilis* is being investigated in strawberry fields by using outside crop rows as predator nurseries. As soon as a few leaves appear on new plants, twospotted mites are seeded onto a 10-foot (3-m) section of an outside row of strawberries upwind from the rest of the crop. Mite populations are carefully monitored, and when they increase, about five *P. persimilis* are released onto each plant. Another 10-foot-long twospotted mite nursery is then established near the crop's center to attract the predatory mites. If careful monitoring finds that predators have not dispersed into the crop's center, leaves containing predatory mites from the original nursery at the edge of the field are clipped and placed in the center nursery or spread to other locations

where twospotted mite populations appear without predators (Strand 1994).

■

MONITORING AND COLLECTING NATURAL ENEMIES

Survey or monitor crops or plants on a regular basis to check for proper cultural care, numbers of pests, and amount of damage. Good monitoring and record keeping tell you when, where, and what kind of natural enemies are present and give at least a general idea of their relative abundance. Knowing that natural enemy populations are high can often allow you to delay or forgo treating a pest population that would otherwise require control action if natural enemies were less abundant. For instance, treatment thresholds for spotted alfalfa aphid are much higher when specific ratios of lady beetles to aphids are present (Summers, Hagen, and Stern 1996). Treatment for citrus red mite is generally not needed if at least one predatory mite is present for every one to two red mites (Grafton-Cardwell et al. 1996). In processing tomatoes, growers can count the ratio of parasitized (black) to unparasitized (white) tomato fruitworm eggs to determine whether pesticide application is warranted (Toscano, Zalom, and Trumble 1995). In addition to allowing use of treatment thresholds based on natural enemies (see table 1-1), monitoring helps time activities (such as beneficial insect releases) that enhance biological control and avoid activities that can damage natural enemies (such as spraying pesticide at the wrong time). Also, because insect development depends on temperature, monitoring degree-days can be used to predict when specific life stages of certain pests and natural enemies are most abundant (Ascerno 1991, Gelernter 1996). Local temperature data and easy-to-use degree-day models are available through various World Wide Web internet sites, including that of the UC IPM Project Internet site listed in the resources at the back of this book.

Proper identification of pests and natural enemies is an essential first step in using biological control. Learn to recognize parasites and predators and their activity as discussed in chapters 7 and 8. Predators of major pests can often be identified based on their appearance by using general references such as UC IPM manuals. Adult parasitic wasps and flies are more difficult to identify and only a few specialists can identify the species of these adult parasites. Therefore, in order to identify parasites, it is important to collect parasitized pests and hold them in containers until the adult parasites emerge. Take organisms that you can't identify to a local Cooperative Extension or county agriculture department office. If they still cannot be identified, ask where they can be mailed to get a proper identification.

After identifying the organisms, learn about their biology. If pests are abundant but no natural enemies are apparent, learn what natural enemies attack those pests and whether anything can be done to enhance biological control.

Most assessments of natural enemy activity include a pest-to-natural-enemy ratio or assessment of the percentage of hosts parasitized and killed. Therefore, hosts as well as natural enemies must be systematically monitored. Regularly inspect a number of leaves, shoots, branches, or terminals or use the specialized techniques described below. Select appropriate methods based on knowledge about pest and natural enemy biology and the goals of your program; consult resources like UC IPM publications and Cooperative Extension advisors. Keep a written description of your monitoring and inspection methods and follow these every time you monitor. Because monitoring also determines whether pest or natural enemy populations or damage are increasing or decreasing, it is important to monitor the same way every time to make results comparable among monitoring dates.

Keep good written records. Write down the date, specific location, number of plants or plant parts (samples) inspected, and the results or counts from your monitoring. Record what

Good monitoring of this growing area includes yellow sticky traps for flying insects and a humidity sensor (the white cylinder) connected to a computerized environmental monitoring and control system. Container soil moisture, light intensity, and temperature are also monitored with sensors not shown here.

Use a 10 power hand lens to detect and identify mites and small insects. Hold the lens near your eye and move the object being viewed close until it is in focus.

management action you took and when you took it, as well as environmental conditions (such as weather) or plant cultural conditions that might affect pests and their biological control. It can be very helpful to enter data into a spreadsheet and to use computerized databases and graphical display to summarize and interpret information.

The most useful method for monitoring natural enemies is to inspect plants for their presence or evidence of their activity, such as parasitized and killed

■ FIGURE 6-5. An aspirator is used to collect tiny or delicate insects. Two bent copper tubes are inserted through a cork that fits snugly into a vial. Screen is placed over the inside end of one tube and a flexible hose is attached to the outside end of that tube. By sucking through the hose, insects near the outside end of the other tube are drawn into the vial and captured. Similar devices filled with alcohol are available for collecting insects and preserving them in the field. Consult the list of suppliers at the end of this book for commercial distributors of insect aspirators. *Source:* Anonymous (1952).

hosts. Many specialized techniques for pest monitoring (table 6-4) can also be used to monitor natural enemies. However, with many of these techniques it can be difficult to relate the natural enemies to a specific pest. Tools such as a 10x hand lens, beating trays, containers, and an aspirator for sucking tiny and delicate creatures into a vial (fig. 6-5), and resource materials such as this book, are helpful.

TABLE 6-4. Monitoring methods for some invertebrate pests and natural enemies.

METHOD	NATURAL ENEMIES	PESTS
visual inspection of plant parts	most species that feed in the open and evidence of parasitism and predation (monitoring may require a hand lens or other magnifier)	most exposed-feeding species and their damage
shaking plants, branch beating, or tapping containers over a collecting surface, such as clipboard with a white sheet of paper	adults and larvae or nymphs of easily dislodged species, including predaceous bugs, lacewings, lady beetles, predaceous mites, adult parasites	bugs, leaf beetles, leafhoppers, mites, non-webbing caterpillars, psyllids, spider mites, thrips, and adults of weevils, whiteflies, and other easily dislodged pests
sticky traps	adult parasitic flies and wasps	adults including fungus gnats, leafminers, psyllids, shore flies, thrips, whiteflies, and winged aphids
pheromone-baited traps	certain parasite adults are attracted to their host's pheromone	many moths, certain beetles, and males of some scale insects
black light or visible light traps	adult lacewings and some other night-flying beneficials	night-flying adults of moths, some beetles (e.g., chafers, white grubs, other scarabs, some leaf beetles), and certain other pests
breathing on infested plant parts	predatory mites will move more quickly than plant-feeding mites when you disturb them by blowing on mite-infested foliage	because of the carbon dioxide in your breath, thrips hidden in buds or terminals will be induced to move if you slowly exhale on infested plant parts
pitfall traps, with or without a bait attractant	predaceous ground beetles, certain spiders	adult weevils
trap board	decollate snail	adult weevils, slugs, snails
host collection and rearing	species that feed inside of their hosts (only the adult stage of parasites and many other insect species can be positively identified and the host must usually be definitely known)	
host dissection	parasite larvae detected by pulling the host caterpillar apart; armored scale parasites observed by examining beneath the scale (routinely used in scientific examination of various hosts for internal parasites)	
degree-day monitoring	many pests and some beneficial species for which researchers have determined temperature development thresholds and rates	
soil drench or flushes using soap, or in some situations plain water	relatively mobile species in soil or other hidden places, including centipedes, millipedes, pyrethrum, symphylans, and larvae of fungus gnats and shore flies; thrips and possibly other species in buds, may also be flushed with pyrethrum	
timed counts	individuals of exposed beneficial and pest species (e.g., lady beetles, caterpillars) or certain types of damage (rolled leaves) that are relatively large and obvious but occur at relatively low density so they are not observed faster than they can be counted	

CHAPTER SEVEN

PARASITES OF ARTHROPODS

Magnifying glass signifies that subject is less than 1 mm.

PARASITES ARE SMALLER than their hosts and develop inside or are attached to the outside of the host's body. Often only the immature stage of the parasite feeds on the host. In contrast to predators, where each individual kills many prey, each immature parasite (more precisely called a parasitoid) typically kills only one host individual during its development (fig. 7-1). Before killing it, parasites often alter their hosts' appearance, behavior, development, feeding, and reproduction.

Because most parasites are small and species spend much of their lives hidden within their hosts, many people are unaware of how common and important parasites are in the control of insects. Almost all pest insects have at least one parasite species that attacks them; most have several. Many parasites are quite specific and will parasitize only one or a few very closely related host species. This host specificity can make parasites more effective natural enemies than predators that graze on a variety of species.

Most parasites of insects are either wasps (order Hymenoptera) or flies (Diptera). Unusual insects in the order Strepsiptera and a few species of beetles and moths are also parasites, but these are not discussed here. Parasites are generally categorized as internal or external. Internal parasites (endoparasites) develop inside their hosts. External parasites (ectoparasites) develop attached to the outside of their hosts, generally inserting only their mouth parts to feed inside.

The adult female parasite typically is adapted to lay her egg (or sometimes a larva) in, on, or near only one host life stage (egg, larva, pupa, or adult). Some parasites attack and emerge from the same host life stage. For example, a pupal parasite lays its egg, feeds as an immature, and emerges as an adult entirely during the host's pupal stage. In other species, one life stage of the host is suitable for attack, but the parasite completes its development and kills the host during a later stage. For example, a larval-adult parasite lays its egg in a larva, but the host is not killed and the mature parasite does not emerge until the host has changed into an adult.

Before being able to mature their eggs and deposit them, females of *Aphytis, Cotesia,* and certain other parasitic wasps and flies require a protein and carbohydrate meal, which they usually obtain by feeding on their hosts. Pest mortality resulting from host feeding by this type of parasite (synovigenic) can be an easily overlooked but important source of biological control

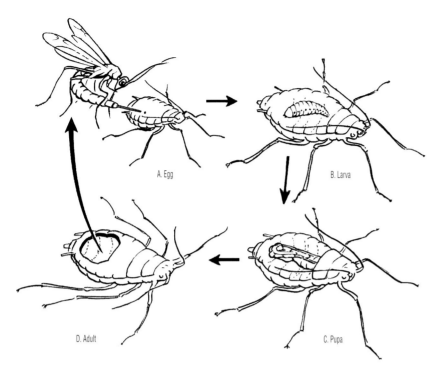

A. Egg

B. Larva

D. Adult

C. Pupa

■ FIGURE 7-1. In many cases, only the immature stage of a parasite feeds on the host. Each immature parasite (often called a parasitoid) kills only one host individual during its development, as illustrated here by the life cycle of a species that attacks aphids. A: An adult parasite lays an egg inside a live aphid. B: The egg hatches into a parasite larva that grows as it feeds on the aphid's insides. C: After killing the aphid, the parasite pupates. D: An adult wasp emerges from the dead aphid, then flies off to find and parasitize other aphids.

An adult *Trichgramma pretiosum* wasp laying its egg inside a bollworm egg. This parasite is about the size of the period at the end of this sentence.

in addition to the host mortality caused by parasitism. Increased host feeding also increases the longevity, host-searching, and egg-laying ability of female parasites. Females of other parasite groups contain fully mature eggs when they emerge from their pupal stage (proovigenic) or they lay eggs that can complete their maturity within the host before hatching.

In some species, only one parasite larva completes development within each host individual (solitary parasites). In other species, multiple eggs can be laid and survive so that many adult parasites develop from a single host (gregarious parasites). Less commonly, a single parasite egg develops into many individuals (polyembryonic parasites). Sometimes parasites themselves are attacked

and killed by other parasite species, called secondary parasites or hyperparasites. Successful parasitism involves five steps: locating the host's habitat, finding the host, accepting (parasitizing) the host, developing in or on the host, and regulating host development.

■

RECOGNIZING PARASITISM

Parasitism is often overlooked because parasites are less familiar than predators and parasitized hosts can superficially appear to be healthy insects. Specific knowledge and careful examination allows parasites to be recognized, and being able to identify and distinguish parasitized hosts is critical in evaluating the activities of parasites. Parasitized hosts often change color because their normal development or metabolism is disrupted or because the host has been killed. A color change also may be caused by the immature parasite itself being visible through the host's outer surface. Parasitized hosts may be mummified so that their outer surfaces harden and remain unusually well preserved well after the host dies and the parasite departs. Excrement from the larval parasite (meconia) is left behind and may be observed inside the host mummy, cocoon, or mine.

Parasite adults may be observed searching infested plants and touching potential hosts with their antennae, mouthparts, or ovipositor. Parasite eggs or the larvae of external parasites may be visible when attached to the host. Many parasites pupate externally, and their pupae or cocoons may be found around or beneath host plants or on plant parts near their dead hosts.

Emergence holes left by adult parasites are often visible in the surface of the dead hosts. The size, shape, number, and location of emergence holes may help to identify the parasite species. For example, emergence holes of parasites of many scales are found in different areas of the dead scale's cover than the emergence holes of hyperparasites (parasites of parasites). Emergence holes made by some parasite species are round and relatively smooth-edged, unlike irregular holes made by the plant-feeding insect that chews its way out of its egg or pupal case. For example, parasites emerging from whitefly pupae leave a round, smooth-edged hole while the adult whitefly itself leaves a ragged, often T-shaped emergence hole. Sometimes the situation is reversed, as with stink bug nymphs, which leave round emergence holes in eggs, unlike the jagged-edge holes of their natural enemies, wasps parasitizing stink bug eggs. Knowing the type of emergence holes made by the pests and beneficial species in your plants will help you distinguish among them.

Unusual host behavior may indicate parasitism. For example, once they are parasitized, some hosts will leave the plant part where they normally feed and move to another location. Where there is evidence of parasitism, dissecting apparently healthy insects may reveal that they contain immature developing parasites.

PARASITIC WASPS

The order Hymenoptera (wasps) includes more parasites than any other insect order. Thousands of different species of parasitic wasps occur in over 40 families. Most parasitic Hymenoptera are small to minute wasps that generally do not sting people. These tiny wasps are very diverse in appearance, biology, and the hosts that they exploit. Most species of insects are attacked by one or more parasitic wasp species during one or more of their life stages.

Table 7-1 describes some common groups of parasitic wasps, and figure 7-2 shows characteristics used to identify wasps. Using this information and more detailed references may allow nonspecialists to identify some parasitic wasps to the family level (Goulet and Huber 1993, Grissell and Schauff 1990) and possibly even to the level of genus (Marsh, Shaw, and Wharton 1987) or

Coccophagus lecanii parasites developing inside caused some of these European fruit lecanium scale nymphs to turn black. Emergence holes left by adult parasites are visible in the outer surface of some dead scales. Dissecting apparently healthy (brownish) scales would reveal that some contain parasites, which have not yet developed enough to discolor their host.

species (Schauff, Evans, and Heraty 1996). However, the species of most parasites can reliably be identified only by experts. Knowing the host that the parasite emerged from and the identifying characteristics of the parasitized host may also allow nonspecialists to identify some common parasites fairly accurately. Books that catalog hosts, such as *Catalog of Hymenoptera North of Mexico* (Krombein, Hurd, and Smith 1979) and well-illustrated publications like *Hymenoptera of the World* (Goulet and Huber 1993) are also very helpful.

Some important parasitic wasps are discussed here in sections organized according to the type of host or insect life stage they attack. Hundreds of other parasite species not illustrated here also are important in pest control in the United States. For other information and photographs of important parasites

in the crops or situations of interest to you, refer to crop-specific publications (such as UC IPM manuals), the references at the end of this book, and other illustrated sources such as university World Wide Web sites accessible over the Internet (see the resources listed at the end of this book).

EGG PARASITES

Wasps in the families Mymaridae, Scelionidae, and Trichogrammatidae parasitize insect eggs exclusively. Egg parasites also occur in some other families, including Eulophidae and Encyrtidae. Most egg parasites are about $\frac{1}{50}$ to $\frac{1}{16}$ inch (0.5–1.5 mm) long; some species are the smallest of all insects, with mature adults about $\frac{1}{125}$ inch (0.2 mm) long. Because these tiny parasites are easily overlooked and difficult to

TABLE 7-1.   Selected families of parasitic wasps.

Aphelinidae*

A

Hosts include aphids, mealybugs, psyllids, scales, and whiteflies. A diverse group of external or internal and primary or secondary parasites. Some males develop as parasites of females. Adults are usually ⅟₂₅ inch (1 mm) long or less. Genera include *Aphelinus, Aphytis, Coccophagus, Encarsia, Eretmocerus,* and *Prospaltella* (=*Encarsia*). One of the most important groups, with about 1,000 known species.

Aphidiidae

B

Internal parasites of aphids. Aphids parasitized by aphidiids typically form tan or golden mummies, unlike aphids parasitized by aphelinids, which usually turn blackish. Small wasps, sometimes included in family Braconidae. Genera include *Aphidius, Lysiphlebus, Praon,* and *Trioxys.*

Braconidae

C

Hosts include larvae of beetles, caterpillars, flies, and sawflies. Most larvae are internal parasites, but many emerge to pupate outside their dead hosts. Adults usually are less than ½ inch (13 mm) long with a slender abdomen longer than the head and thorax combined. Genera include *Bracon, Chelonus, Cotesia* (=*Apanteles*), and *Opius.* Over 1,000 known species.

Chalcididae*

D

Mostly internal or external parasites of Diptera or Lepidoptera larvae or pupae, with some species attacking beetles or other wasps. Adults are robust, have a greatly enlarged hind femora (leg segment), and are often dark and shiny. About 1,500 species are known.

Encyrtidae*

E

Internal parasites of ticks and various insect eggs, larvae, or pupae, including beetles, bugs, moths, mealybugs, and scales. Adults are usually less than ⅟₁₂ inch (2 mm) long. Genera include *Anagyrus, Comperiella, Copidosoma, Encyrtus, Leptomastix, Metaphycus, Pentalitomastix,* and *Psyllaephagus.* Over 3,000 species.

Eulophidae*

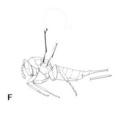

F

Internal or external parasites of eggs, larvae, or pupae of flies, moths, and other wasps; also parasitize mites, spiders, scale insects, and thrips. Adults are small but usually ⅟₂₅ inch (1 mm) or larger. Genera include *Aprostocetus, Chrysocharis, Diglyphus, Oomyzus, Tetrastichus,* and *Tamarixia.* About 3,400 species are known.

---

study, the specific hosts and correct identity of many species are uncertain. Although many publications discuss wasps in the genus *Trichogramma,* most reports are difficult to interpret. Few entomologists can properly distinguish the species in this genus and the incorrect species name is often used.

Moth Egg Parasites

Trichogramma spp. (Trichogrammatidae) wasps parasitize eggs of hundreds of different species of pests, especially eggs of moths and butterflies. Some *Trichogramma* spp. show little or no preference for eggs of one particular species; these wasps may be habitat specific, attacking eggs of virtually any insect present in that setting.

Natural populations of *Trichogramma* can occur at high densities and are important in biological control of bollworm or tomato fruitworm infesting tomatoes (Hoffmann et al. 1990, 1991a) and certain other crops. Several *Trichogramma* species are commercially available and are released against pests in forests (Bai, Cobanoğlu, and Smith 1995) and certain field crops (Smith 1996, Wajnberg and Hassan 1994). Reportedly successful releases include *Trichogramma nubilale* to control European corn borer (Losey et al. 1995), *Trichogramma platneri* against amorbia in avocado (Oatman and Platner 1985)

and tomato fruitworm (Toscano, Zalom, and Trumble 1995), and *T. platneri* combined with mating disruption for control of codling moth in apples (Caprile et al. 1996).

Leafhopper Egg Parasites

Leafhoppers suck juices from many plants. *Anagrus* spp. (Mymaridae) are ⅟₁₀₀ to ⅟₂₀ inch (0.3–1.1 mm) long wasps that parasitize eggs of leafhoppers, mealybugs, true bugs, and other insects.

A complex of *Anagrus* spp., including *A. epos,* helps control grape and variegated leafhoppers in vineyards. *Anagrus* spp. also parasitize eggs of the prune leafhopper. Blackberries that

TABLE 7-1. Selected families of parasitic wasps *(continued)*.

Ichneumonidae

External or internal parasites. Larvae or pupae of most types of insects are attacked by one or more species; beetles, caterpillars, and wasps are common hosts. Adults are usually slender (the abdomen is longer than the head and thorax combined) with a long ovipositor and 16 or more antennal segments. Genera include *Bathyplectes, Campoplex, Diadegma, Exochus, Hyposoter, Ophion,* and *Venturia.* Over 3,100 species in North America.

Scelionidae

Internal parasites of spider and insect eggs, especially bugs and moths. Adults are usually less than ½s inch (1 mm) long. Females of some scelionids attach themselves to the female of the host and ride on it; when the host lays its eggs, the scelionid leaves the host and attacks the eggs (one insect attaching to another for transportation is called *phoresy*). *Telenomus* and *Trissolcus* are important genera. About 300 species in United States and Canada.

Mymaridae*

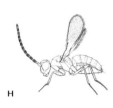

Internal parasites of insect eggs, including beetles, flies, grasshoppers, leafhoppers, and true bugs. Adults are less than ½s inch (1 mm) long and are distinguished by their stalked, narrow, elongate hind wings. Both wings are often fringed with fine hairs. Genera include *Anagrus* and *Anaphes.* Over 1,300 known species.

Trichogrammatidae*

Internal parasites of insect eggs. *Trichogramma* spp. attack many types of insects, especially moths and sawflies. Some species prefer eggs of certain types of hosts while others prefer certain habitats and may be able to parasitize almost any insect egg in that environment. Adults are ½s inch (1 mm) or less and often have wing hairs (setae) arranged in rows. About 650 species.

Pteromalidae*

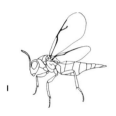

Mostly parasites of beetles, flies, and other wasps. Many are secondary parasites. Biology and appearance varies greatly among species. Adults are usually a little more than ½s inch (1 mm) long and black or metallic-green or bronze. Genera include *Dibrachys, Perilampus, Pteromalus,* and *Scutellista.* Over 3,000 species.

*Families Aphelinidae, Chalcididae, Encyrtidae, Eulophidae, Mymaridae, Pteromalidae, and Trichogrammatidae are among those in the superfamily Chalcidoidea. The term *chalcid* commonly is used to refer both to family Chalcididae and to any wasp in other families within Chalcidoidea. Specialists sometimes disagree on insects' classification. For example, some specialists consider aphelinids to be subfamily Aphelininae within family Encyrtidae, so they may refer to aphelinids as encyrtids.

Sources: Borror, De Long, and Triplehorn 1981; Daly, Doyen, and Ehrlich 1978; Goulet and Huber 1993; Grissell and Schauff 1990 (A–C, E–F, H, J–K *from* Goulet and Huber 1993, J by S. Rigby, reproduced with the permission of the Minister of Public Works and Government Services Canada, 1997; D, I *from* Grissell and Schauff 1990; G *from* Townes and Townes 1960).

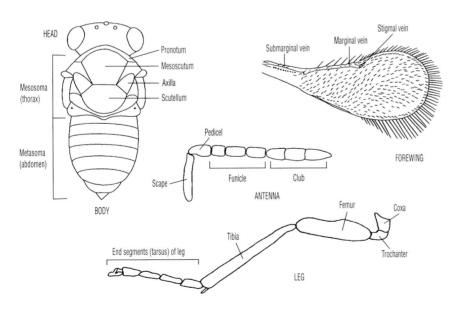

■ FIGURE 7-2. Principal characteristics for distinguishing parasitic wasps as illustrated by an *Encarsia* sp. Although the species of parasites can usually be distinguished only by an expert, species of *Encarsia* parasitizing whiteflies in North America can be identified to species by some nonspecialists with access to a compound microscope and the patience to carefully spread tiny specimens onto slides so that identifying characters are visible. Keys to the six genera of whitefly parasites and to the *Eretmocerus* spp. in the continental United States are available in Rose and Zolnerowich (1997). Adapted from Schauff, Evans, and Heraty (1996).

Parasitized eggs that contain a *Trichogramma* sp. typically darken, such as the central and right corn earworm eggs shown here. Because their life cycle from egg to adult is about 7 to 10 days, these parasites have many more generations than their hosts, so populations of these natural enemies can increase rapidly.

LENGTH
Trissolcus basalis ovipositing in an egg cluster of the southern green stink bug. Parasitized eggs darken; if parasites have emerged, the emergence holes will be irregular as opposed to round holes caused by stink bugs pushing off the top of the cap to emerge. In some areas, this scelionid wasp kills up to 90 percent of the stink bug eggs infesting tomatoes (Hoffmann et al. 1991a, Weber et al. 1996).

LENGTH
One *Hyposoter fugitivus* is pupating inside each of these dead redhumped caterpillars.

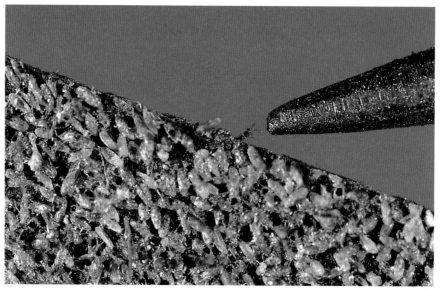

These commercial *T. pretiosum* are shipped as parasitized moth eggs glued to a card. The pencil points to a recently emerged adult parasite. Because they attack only eggs, any releases of *Trichogramma* must be made early and throughout the moth's egg-laying cycle. Pheromone-baited traps are a good tool for timing *Trichogramma* releases, because pest egg laying coincides with periods when flying moths can be trapped.

This normally pale-colored grape leafhopper egg visible within leaf tissue has turned reddish because it is parasitized by an *Anagrus* sp.

host leafhoppers and certain other plants have been recommended as alternate hosts for leafhopper parasites near grapes; more recent research indicates that these other leafhoppers are attacked by similar looking but different parasite species that will not move to attack grape leafhoppers. However, these tiny fringe-winged mymarid parasites migrate primarily by floating with the wind, so nearby plantings can provide a windbreak that increases colonization of grapes by windborne *Anagrus* spp. (Corbett and Rosenheim 1996; Murphy, Rosenheim, and Granett 1996).

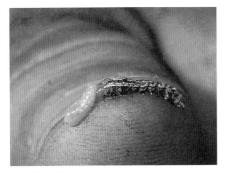

A green *Hyposoter exiguae* larva will pop out if you pull apart a parasitized armyworm.

A white, speckled cocoon of *Hyposoter exiguae* is attached beneath the skin of an armyworm that it killed.

This shiny, white parasite larva was exposed by grasping the caterpillar at each end and pulling it apart. Late immature stages of *Cotesia medicaginis* cause parasitized alfalfa caterpillars to appear lighter than normal, somewhat shiny rather than velvety on the surface, and swollen toward the rear. The parasitized host dies before it reaches ½ inch (12 mm) long.

CATERPILLAR PARASITES

Caterpillars (the larvae of moths and butterflies, order Lepidoptera), are commonly parasitized by braconid and ichneumonid wasps. Lepidoptera pupae are often parasitized by chalcids, and butterfly eggs are frequently parasitized by trichogrammatids.

Hyposoter

At least 27 *Hyposoter* spp. (Ichneumonidae) in the United States attack dozens of different caterpillars. *Hyposoter fugitivus* and *H. exiguae* are among the most common of these larval endoparasites. Hosts of *Hyposoter* spp. include armyworms, cabbage looper, California orangedog, fall webworm, oakworms, redhumped caterpillar, tent caterpillars, tomato fruitworm, and tussock moths.

Adult *Hyposoter* are ¼ to ½ inch (6–12 mm) long and are black with lighter-colored legs. Although females prefer to parasitize young larvae, they will lay one egg in almost any instar. The dead host shrinks and becomes hard and brittle. The parasite then spins a cocoon inside the larval skin, or the caterpillar skin splits and the parasite emerges and pupates outside the host. The dead host, parasite pupae, or both often adhere to the host plant singly or in small clusters. Parasite development from oviposition until adult emergence typically takes about 1 month.

Cotesia

Over 200 species of wasps in the genus *Cotesia* (=*Apanteles*) (Braconidae) occur in North America; most are parasites of exposed-feeding Lepidoptera. Common hosts include alfalfa caterpillar, armyworms, conifer budworms, orange tortrix, gypsy moth, imported cabbageworm, loopers, redhumped caterpillar, tent caterpillars, and tussock moths.

Bracon

About 100 *Bracon* species (Braconidae) occur in the United States. Most are external parasites of Lepidoptera or Coleoptera larvae; a few parasitize flies or sawflies. Hosts include conifer budworms, grape leaffolder, pink bollworm, pine tip moths, and tent caterpillars.

Cotesia medicaginis emerged to pupate in the open and formed this white cocoon after the parasite larva fed inside and killed a host caterpillar.

Larvae of *Bracon cushmani*, a gregarious external parasite feeding on a grape leaffolder larva.

Several pale, oblong eggs of *Bracon cushmani* laid externally on a grape leaffolder larva.

Larvae of *Goniozus legneri* feed attached to the outside of navel orangeworms. Several parasite larvae (top) have been removed to reveal their shriveled, dead host. The number of parasites per host ranges from a few to about 16 (Gordh, Woolley, and Medved 1983).

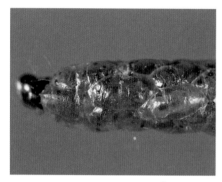

LENGTH ⊢⊣ Immature *Pentalitomastix plethoricus* polyembryonic wasps are visible through the outer skin of this navel orangeworm. The adult *Pentalitomastix* is ¹⁄₂₅ inch (1 mm) long and is shiny purple to dark brown or black. As with *Copidosoma* spp., each female lays a single egg in a navel orangeworm egg. About 500 individual wasps develop from one parasite egg, consuming and entirely filling the dead host's body.

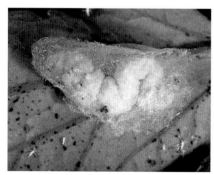

This looper parasitized by *Copidosoma truncatellum* curls into an S shape after spinning its cocoon. The looper fails to pupate into a moth, and as many as 500 small wasps emerge from the dead larva.

LENGTH ⊢⊣ This female of *Goniozus legneri*, which is ⅙ inch (4 mm) long, paralyzes navel orangeworms such as this by stinging them near the head and injecting venom. The wasp then walks repeatedly back and forth along the larva, occasionally biting it to make sure it is paralyzed and sometimes stinging it repeatedly. Usually within about 30 minutes, the wasp lays several pale, oblong eggs on the host. The stung caterpillar is permanently unable to move, but it remains alive for days until consumed by the parasite larvae.

Copidosoma

About a dozen *Copidosoma* spp. (Encyrtidae) in the United States are endoparasites of Lepidoptera. *Copidosoma truncatellum* parasitizes armyworms and loopers. Adults are ¹⁄₂₅ inch (1 mm) long and are shiny purple to dark brown or black. Females lay one or a few eggs in a caterpillar egg. After the host caterpillar emerges from its egg, the parasite egg develops into a mass of tissue. After the host larva develops into a mature larva, it is killed as this mass of tissue develops into hundreds of individual parasites. This parasite species is called polyembryonic because many individual wasps originate from a single egg. The wasps, which are either all males or all females, complete their development and emerge. Parasite development from egg laying to adult emergence takes from 3 weeks at 85°F (30°C) to 4 months at 60°F (16°C).

Navel Orangeworm Parasites

Navel orangeworm (NOW) is the most important insect pest of almonds.

It also feeds on dried, decaying, and damaged fruit, legume seed pods, pomegranate, stone fruits, and nuts such as pistachio and walnut. *Pentalitomastix plethoricus* (=*Copidosomopsis plethorica*) (Encyrtidae) and *Goniozus legneri* (Bethylidae) were introduced from Mexico and South America, respectively, to help control this pest (Legner and Gordh 1992).

Removing old almonds (mummy nuts) left on trees after harvest is a highly effective cultural control that eliminates overwintering navel orangeworms. Removing mummy nuts apparently does not reduce biological control of navel orangeworm because mummy nuts contain few if any overwintering parasites. Commercial parasites may be available for release. Any parasite releases must be supplemented with cultural and other practices to conserve parasites and control the navel orangeworm, as detailed in *BIOS for Almonds* (Anonymous 1995b), *Integrated Pest Management for Almonds* (Anonymous 1985a), and *Almond Pest Management Guidelines: Insects* (Zalom et al. 1996).

BEETLE PARASITES

There are more species of beetles (order Coleoptera) than any other insect. Beetles are attacked by many different parasites, including Braconidae, Eulophidae, Ichneumonidae, and Pteromalidae.

Alfalfa Weevil Parasites

The alfalfa weevil was accidentally introduced into North America from the Old World in the early 1900s. It is a pest of alfalfa throughout the United States, feeding on young shoots and defoliating plants. At least seven parasite species have been introduced from Europe to control it, including larval-prepupal parasites (*Bathyplectes curculionis* and *B. anurus*, Ichneumonidae), adult and adult-larval parasites (*Microctonus aethiopoides* and *M. colesi*, Braconidae), and the larval parasite

Many insects can defend themselves against parasitism through physiological and behavioral mechanisms. Parasites have in turn evolved ways to overcome these defenses, as illustrated by the alfalfa weevil.

There are two identical-looking strains of alfalfa weevil, one in the western United States and one in the East. Each strain has a different defense against parasitism. In the eastern United States, weevils use encapsulation to avoid parasitism by *B. curculionis*. Soon after the parasite egg is laid, the host's blood cells (hemocytes) aggregate and enclose the parasite in a sheath. This capsule suffocates and eventually decomposes the young parasite. Because of this encapsulation in the East, *B. curculionis* is less effective and *B. anurus*, which is not encapsulated, is the more effective parasite.

Since the western alfalfa weevil strain exhibits a low rate of encapsulation (possibly because parasites have evolved to avoid this), *B. curculionis* is very effective in many areas of the West. However, western weevils have a different defense against another solitary adult endoparasite, *Microctonus aethiopoides* (Braconidae). Western weevils contain rickettsia, bacterialike microorganisms that often inhabit arthropods. These rickettsia kill *M. aethiopoides* parasites. Apparently because of the rickettsia, *M. aethiopoides* is not important in the West, even though it helps control alfalfa weevil in the East (Maund and Hsiao 1991).

Oomyzus (=*Tetrastichus*) *incertus* (Eulophidae). Along with the entomogenous fungus *Zoophthora phytonomi*, which attacks larvae and pupae, these natural enemies effectively control alfalfa weevil in the eastern United States but are less effective in the West. The Egyptian alfalfa weevil, a similar-looking pest species that also occurs in the West, is not controlled well by parasites.

Adult *Bathyplectes* are robust, black wasps that are about 1/12 inch (2 mm) long. Both *Bathyplectes* spp. are solitary endoparasites that prefer to parasitize early-instar weevil larvae. Parasite larvae emerge from dead weevil larvae or pupae and the parasite forms a brown, white-banded pupal case, often inside the weevil's own coarsely woven, netlike, white cocoon. *Bathyplectes curculionis* occurs throughout the United States and has two generations a year. Second-generation parasites that emerge the same year form a light brown pupal case, in contrast with the dark brown pupae of overwintering parasites. *Bathyplectes anurus* occurs primarily in the eastern United States and has one generation a year.

Lady Beetle Parasite

Not all parasites are beneficial. *Dinocampus* (=*Perilitus*) *coccinellae* (Braconidae) attacks more than 40 species of aphid-feeding lady beetles (Obrycki 1989). Female *D. coccinellae* actively pursue and oviposit in adult

A brown, white-banded *Bathyplectes* pupa within the coarsely woven silk made by the alfalfa weevil pupa it killed. Pupae of some parasites can defend themselves against secondary parasites. When a parasite approaches, *Bathyplectes anurus* suddenly straightens itself within its pupal case, causing it to jump and fall to the ground, where it will likely escape the parasite. Pupae of a related parasite, *Bathyplectes curculionis*, do not jump and the white band around them is not raised, while the white band around *B. anurus* is raised.

LENGTH

LENGTH

An adult *Dinocampus coccinellae* and its silken cocoon next to a convergent lady beetle, which the parasite killed.

beetles and occasionally in larvae. Larvae feed inside and develop through four instars, then exit the host and spin a cocoon. *Dinocampus coccinellae* generally attacks the larger species of lady beetles, reducing their effectiveness as biological control agents. Studies have found that about 10 to 30 percent of convergent lady beetles are parasitized by *D. coccinellae*.

■

Eucalyptus Beetle Parasites

Eucalyptus pests, including the blue gum psyllid and two beetles, have been targets of parasite importation programs conducted by the University of California. The eucalyptus snout beetle has been controlled through introduction of an egg parasite (*Anaphes nitens*, Mymaridae); this serious defoliator of eucalyptus from Australia was accidentally introduced into Southern California in the 1990s (Cowles and Downer 1995) and temporarily defoliated entire trees until brought under complete biological control. Several parasites have also been introduced to help control the eucalyptus longhorned borer (Hanks et al. 1995; Hanks, Paine, and Miller 1996), which has killed many eucalyptus trees throughout California.

■

SCALE PARASITES

Parasitic wasps such as *Aphytis*, *Coccophagus*, *Encarsia*, and *Metaphycus* spp. are the most important natural enemies of many scale insects (table 7-2). Even scale species with effective parasites can be temporarily abundant until natural enemies colonize the scales and reproduce. If scale outbreaks persist because natural enemies have been disrupted (such as by ants), measures to conserve and enhance biological control may be warranted (see chapter 6).

Parasites and certain other natural enemies are commercially available for release against certain scale species (see tables 6-3 and 7-2). Releases of *Aphytis melinus* are recommended in citrus to control California red scale, but for most scale insects there is little information on the effectiveness of releasing natural enemies to provide control. If releases are planned, follow the guidelines for releasing natural enemies effectively in chapter 6.

TABLE 7-2. Selected scales in California that can be partially to completely controlled by introduced natural enemies.

COMMON NAME OF SCALE	SCIENTIFIC NAME	TYPE	NATURAL ENEMIES
black scale	*Saissetia oleae*	S	*Metaphycus helvolus**, *Metaphycus* spp. parasitic wasps
brown soft scale	*Coccus hesperidum*	S	*Metaphycus luteolus*, *Metaphycus* spp., *Microterys* sp.* parasitic wasps
California red scale	*Aonidiella aurantii*	A	*Aphytis melinus**, *Aphytis lingnanensis*, *Comperiella bifasciata*, *Encarsia perniciosi* parasitic wasps
cottony cushion scale	*Icerya purchasi*	M	*Rodolia cardinalis* lady beetle, *Cryptochaetum iceryae* parasitic fly
dictyospermum scale	*Chrysomphalus dictyospermi*	A	*Aphytis melinus** parasitic wasp
fig scale	*Lepidosaphes conchiformis*	A	*Aphytis mytilaspidis* parasitic wasp
green shield scale†	*Pulvinaria psidii*	S	*Cryptolaemus montrouzieri** mealybug destroyer lady beetle *Metaphycus* spp. parasitic wasps
hemispherical scale	*Saissetia coffeae*	S	*Metaphycus helvolus**, *Metaphycus* spp. parasitic wasps
ice plant scales	*Pulvinaria delottoi*, *Pulvinariella mesembryanthemi*	S	*Metaphycus funicularis*, *M. stramineus* parasitic wasps *Exochomus flavipes* lady beetle
nigra scale	*Parasaissetia nigra*	S	*Metaphycus helvolus** parasitic wasp
obscure scale	*Melanaspis obscura*	A	*Encarsia aurantii* parasitic wasp
olive scale	*Parlatoria oleae*	A	*Aphytis maculicornis*, *Coccophagoides utilis* parasitic wasps
purple scale	*Lepidosaphes beckii*	A	*Aphytis lepidosaphes* parasitic wasp
San Jose scale	*Quadraspidiotus perniciosus*	A	*Encarsia* (=*Prospaltella*) *perniciosi* parasitic wasp
walnut scale	*Quadraspidiotus juglansregiae*	A	*Aphytis*, *Encarsia* (=*Prospaltella*) spp. parasitic wasps *Chilocorus orbus*, *Cybocephalus californicus* beetles
yellow scale	*Aonidiella citrina*	A	*Comperiella bifasciata* parasitic wasp

Note: Scale types are: A (armored, Diaspididae); M (Margarodidae); and S (soft, Coccidae). Consult the Index for order and family names of invertebrates mentioned here.

Sources: Anonymous 1987, 1991b; DeBach 1964; Ehler 1995; Flint et al. 1992; Laing and Hamai 1976.

* Species may be commercially available.

† Green shield scale may be quarantined by the California Department of Food and Agriculture. Although biological control appears to be important in limiting scale populations while growing infested plants, plants may be required to be entirely free of this pest to be marketed, which may require thorough spraying and careful inspection before shipping.

Metaphycus—A Soft Scale Parasite

Although *Metaphycus helvolus* (Encyrtidae) kills several species of soft scales, its most important host is black scale. Black scale is a dark, dome-shaped scale that infests many agricultural and ornamental plants and is attacked by over 50 parasite species (Daane et al. 1991, Lampson and Morse 1992). In addition to black scale, hosts of *M. helvolus* include brown soft scale and European fruit lecanium (table 7-2).

Metaphycus helvolus is a solitary endoparasitic wasp that is ¹⁄₂₅ inch (1 mm) long. The yellowish orange females attack first-instar through early third-instar nymphs. *Metaphycus* kills scales both by parasitization and by piercing scale nymphs and feeding on the exuding contents (host feeding). It is highly effective in areas with mild winters, except when ants are present. Because *M. helvolus* takes several minutes to lay each egg, it is more susceptible to disruption by ants that tend honeydew-producing insects than parasite species that oviposit faster. *Metaphycus helvolus* has several generations per year, whereas black scale has only one or two generations per year.

Commercially available *M. helvolus* can be released when second-instar host scales are common. In black scale, this is the young, flattened, brownish stage when an H-shaped ridge is apparent on the scale's back. If *M. helvolus* is released in citrus, the recommendation is a minimum of 2,000 parasites per acre in late summer or early fall (Grafton-Cardwell et al. 1996). Another commercially available parasite species, *Microterys nietneri* (=*M. flavus*) attacks only brown soft scale. Although natural populations are important outdoors, there is little research on the effectiveness of releasing soft scale parasites.

California Red Scale Parasites

Aphytis melinus (Aphelinidae) is the most important parasite attacking California red scale, a small, flattened, reddish orange scale infesting citrus and

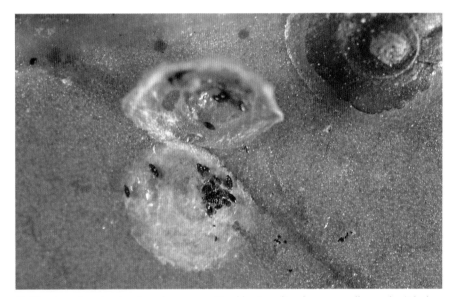

Metaphycus helvolus emerged from a hole in these two second-instar black scale nymphs. A more mature, darker, unparasitized black scale is in the center.

California red scales that have been parasitized by *A. melinus* have a small round exit hole. Here, the cover of a parasitized scale has been lifted and turned upside down, revealing the dried, flat skin of the scale body left by this external parasite. Parasite fecal pellets (meconia) are visible on the citrus leaf. If this scale had been parasitized by *Comperiella bifasciata*, the exit hole in the cover would be larger and more irregular. Because *C. bifasciata* is an internal parasite, underneath the scale cover would be a parchmentlike, bloated scale skin containing meconia.

many ornamental plants. *Life Stages of California Red Scale and Its Parasitoids* (Forster, Luck, and Grafton-Cardwell 1995) vividly illustrates the various scale and parasite stages. *Aphytis melinus* also attacks several species of armored scales, including latania scale and oleander scale.

Female *A. melinus* feed on and oviposit in immature scales, preferring the virgin adult female scale. In California red scale, this preferred virgin female stage has a wide gray margin extending beyond the insect body, and the scale cover and body can be readily separated. The solitary, ectoparasitic larva leaves a flat and dehydrated scale body beneath the scale cover, where the parasite's cast skin and fecal pellets (meconia) may be observed. The tiny yellow adult wasp often chews a small round emergence hole in the scale cover; other female *Aphytis* emerge from beneath the scale cover, so that parasitized scales slough off and the evidence parasitism is removed. The parasite's short life cycle (10 to 20 days)

LENGTH ⊢

The adult male California red scale shown here is easily confused with the scale parasite *Aphytis melinus*. Unlike the parasite, the male scale has long antennae, a dark band around its back, and only one pair of wings.

results in two or three parasite generations for each scale generation.

In addition to the importance of conserving natural populations, releasing commercially available *A. melinus* is effective in citrus if ants that tend honeydew-producing insects are controlled and broad-spectrum pesticides are avoided for all citrus pests. A total of 50,000 to 150,000 parasites per acre per year is typically released. The greatest number of *Aphytis* are released during spring. Parasites are released at two-week intervals beginning in February when temperatures reach 60°F (16°C) and virgin female scales are present (Forster, Luck, and Grafton-Cardwell 1995; Grafton-Cardwell et al. 1996; Moreno and Luck 1992).

Comperiella bifasciata is an important encyrtid that parasitizes California red scale. It also attacks yellow scale and Florida red scale. Adult parasites are black, with two white stripes on the female's head. *Comperiella* will attack hosts at almost any stage, except gravid females (those with eggs). One parasite generation requires about 3 to 6 weeks to develop, with quicker development happening on larger (later instar) hosts and at warmer temperatures.

APHID PARASITES

Common aphid parasites include species in the genera *Aphelinus*, *Aphidius*, *Ephedrus*, *Praon*, and *Trioxys*. These parasites reproduce by laying their eggs in aphids (fig. 7-1). The immature wasp feeds inside the host and kills it, causing the aphid to become slightly puffy or mummified. Tan or golden aphid mummies typically are produced by aphidiid parasites; blackish mummies are usually produced by aphelinids. A round hole can be observed where the adult parasite has chewed its way out of the aphid mummy. Parasites are very important in the naturally occurring biological control of many important aphids. Although a few well-known parasites attack many different aphid species, many parasitize only one or a few closely related species of aphids.

Several aphid predators and parasites are commercially available (tables 6-3 and 7-3); these are sometimes introduced in greenhouses and in interior plantscapes. However, in most situations, there is little research-based information on the effectiveness of introducing predators or parasites for aphid control.

LENGTH ⊢

Diaeretiella rapae is shown here ovipositing in a cabbage aphid, which is one of its primary hosts. This parasite is also commercially available for release against a wide range of aphids, especially *Brachycaudus* and *Myzus* spp.

LENGTH ⊢

An *Aphytis melinus* among its preferred hosts, virgin adult female California red scales.

TABLE 7-3.   Selected commercially available natural enemies of aphids.

NATURAL ENEMY	COMMENTS
aphid midge (*Aphidoletes aphidimyza*)	tiny flies (family Cecidomyiidae) with predaceous larvae (see fig. 8-10); does best in greenhouses or outdoors where humidity is high
aphid parasites (many species, including *Aphidius matricariae* and *Diaeretiella rapae*)	tiny wasps that lay eggs in many species, of aphids (e.g., *Aphis, Brachycaudus,* and *Myzus* spp.); parasitized aphids become tan mummies
convergent lady beetle (*Hippodamia convergens*)	common orange and black-spotted species discussed in chapter 8
green lacewings (*Chrysoperla* spp.)	each larva looks like a tiny alligator and can consume several hundred aphids over its lifetime; because most aphids are eaten by older larvae, there may be a lag time in any control after release until lacewing larvae mature
minute pirate bugs (*Orius* spp.)	*Orius* spp. are released primarily against thrips; natural populations outdoors may feed on aphids, but there is little information on the effectiveness of releasing *Orius* for aphid control

Note: Consult the index for order and family names of the invertebrates mentioned here.

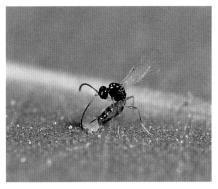

This adult *Trioxys pallidus* has a shiny black head and thorax and a long, slender, yellowish or orangish abdomen. This female is ovipositing into a walnut aphid.

LENGTH

PESTICIDE-RESISTANT NATURAL ENEMIES

Pesticide-resistant strains of certain natural enemies can survive better and provide control even when broad-spectrum pesticides are sprayed against other pests (Hoy 1993; Spollen and Hoy 1992; Spollen, Johnson, and Tabashnik 1995). These pesticide-resistant natural enemies have been developed in the laboratory or evolved naturally and were collected and mass-cultured for sale. Genetic engineering is also being used experimentally to improve the pesticide tolerance of natural enemies (Hoy 1993). Pesticide-resistant strains may be commercially available for several species, including parasitic *Aphytis melinus* and *Trioxys pallidus*, predaceous *Chrysoperla carnea*, and certain phytoseiid mites. However, even if pesticide-resistant natural enemies are present, any pesticides must still be carefully chosen and wisely used. Pesticide-resistant natural enemies are not immune to all doses and types of pesticides, and there may be many other important naturally occurring beneficial species that have no resistance but are needed for pest control.

This *Lysiphlebus testaceipes* emerged from one of these dead cotton aphids by chewing a round hole. Aphids parasitized by wasps in the family Aphidiidae typically form golden or tan mummies such as these.

LENGTH

Walnut Aphid Parasite

Walnut aphid is a small, yellowish insect that feeds along the veins on the underside of walnut leaves. If ants are controlled and broad-spectrum pesticides are avoided, walnut aphid in California can be controlled by an introduced parasite, *Trioxys pallidus* (Aphidiidae). Prior to establishment of this parasite in the late 1960s (van den

Bosch et al. 1979), walnut aphid caused severe crop losses each year.

Trioxys pallidus has several biotypes, as is common with many parasites. Biotypes are populations or strains that have different behaviors and may be different species but are lumped together as one species because they are morphologically indistinguishable. Certain biotypes of *T. pallidus* also

Green peach aphids parasitized by an *Aphelinus* sp. Aphids parasitized by Aphelinidae typically form blackish mummies.

LENGTH

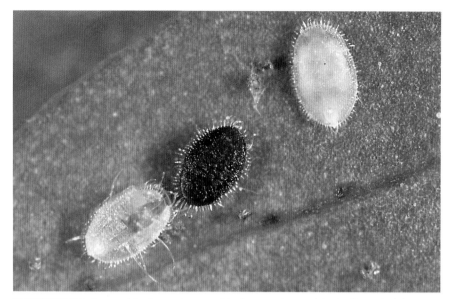

LENGTH ⊢⊣

A parasitized greenhouse whitefly pupa (center) looks like a dark brown or black scale insect. An unparasitized whitefly pupa (right) is whitish or yellowish green. Whitefly nymphs containing young parasites look like unparasitized whiteflies until about 7 to 10 days after egg laying by *E. formosa*, when parasitized scales darken. The ragged or T-shaped slit in the mostly clear, empty pupal skin (left) was caused by an emerging adult whitefly. Whitefly parasites leave a round emergence hole.

attack the filbert aphid and linden aphid (Messing and Aliniazee 1989)

■

Cotton Aphid Parasite

The cotton or melon aphid is usually blackish or dark green, but yellow or whitish forms also occur. Hosts include cotton, coffee, potato, and many ornamentals. *Lysiphlebus testaceipes* (Aphidiidae) is a slender, dark greenish to black parasite that attacks cotton aphid and other aphid species in the genera *Brachycaudus, Macrosiphum,* and *Myzus.*

■

WHITEFLY PARASITES

Most whiteflies in their native habitat are controlled by natural enemies. Several accidentally introduced species are under effective classical biological control (table 7-4). Parasitic wasps, such as *Amitus, Eretmocerus,* and *Encarsia* spp. (fig. 7-3), are important natural enemies of whiteflies.

■

Greenhouse Whitefly Parasite

Greenhouse whitefly infests many plants inside greenhouses and outdoors in warm locations. *Encarsia formosa* (Aphelinidae) is the most important parasite of greenhouse whitefly.

Female *E. formosa* feed on all immature whitefly stages (host feeding) by puncturing whitefly nymphs with their ovipositors and consuming the exuding blood. Females oviposit in third-instar and fourth-instar whiteflies. Eggs hatch into larvae that feed inside and kill the host (fig. 7-3). Greenhouse whitefly nymphs turn dark brown or black (the black scale stage) about 7 to 10 days after being parasitized by *E. formosa.* Like many whitefly parasites, *Encarsia* leaves a round hole and black feces (meconia) in the host remains; a whitefly leaves a ragged or T-shaped emergence hole in its mostly clear or whitish pupal skin.

Biological control of greenhouse whitefly can often be provided in greenhouses by introducing sufficient numbers of commercially available *E. formosa* at the right times. Parasites are usually introduced beginning when whiteflies are first detected. Multiple releases at two-week intervals, continuing until after parasitized whiteflies are common, are sometimes recommended.

In certain greenhouse-grown vegetables, 2 or 3 inoculative releases each growing season may be adequate. Homoptera-tending ants must be controlled, and persistent, broad-spectrum pesticides must be avoided, although certain materials such as abamectin (Zchori-Fein, Roush, and Sanderson 1994) or non-persistent insecticides such as oil or soap may be integrated with releases of *E. formosa.*

Temperatures of about 65°F to 85°F (19°C to 30°C), relative humidities averaging 50 to 80 percent, and adequate lighting (about 650 footcandles or more) are optimum for the development of *E. formosa.* Egg laying declines dramatically or stops during periods of short day length and when temperatures approach or drop below about 55°F (13°C). Prolonged cool, dark conditions can cause *E. formosa* and certain other natural enemies to become locally extinct, which is one reason why *E. formosa* often needs to be reintroduced each season.

■

Silverleaf Whitefly Parasites

Certain strains of *Encarsia formosa* (such as the Beltsville strain) also attack silverleaf whitefly and sweetpotato whitefly (*Bemisia* spp.), although the effectiveness of *E. formosa* in controlling *Bemisia* spp. is debatable (Hoddle and Van Driesche 1996, Parrella et al. 1991). *Eretmocerus* spp. (Rose and Zolnerowich 1997), *Encarsia luteola,* and *Encarsia pergandiella* are silverleaf whitefly parasites occurring naturally in the field. These parasites are adapted to *Bemisia* spp. and some of these species may be commercially available. Naturally occurring biological control is currently not effective in controlling silverleaf whitefly.

■

MEALYBUG PARASITES

Naturally occurring lady beetles, green lacewings, predatory midges, and parasitic wasps provide good control of many mealybug species outdoors unless disrupted by ants or pesticides. Parasites, including *Allotropa, Anagyrus,*

Adult *Encarsia formosa* are tiny wasps with a dark brown to black head and thorax and a bright yellow abdomen.

H
LENGTH

H
LENGTH

Silverleaf whitefly pupae do not darken when parasitized by *Encarsia pergandiella* (top). Orangish parasite fecal matter (meconia) and a round emergence hole remain after *E. pergandiella* emerges (left). When silverleaf whitefly is parasitized by *E. luteola* (right and bottom), the normally clear ridges on the whitefly pupal case darken as the parasite develops. Prior to emergence, the wasp's yellow and dark brownish body becomes visible through the pupal case when viewed with a hand lens. Silverleaf whiteflies parasitized by *E. formosa* look similar to silverleaf whiteflies parasitized by *E. luteola*; they do not have the black pupal case formed when *E. formosa* parasitizes greenhouse whitefly.

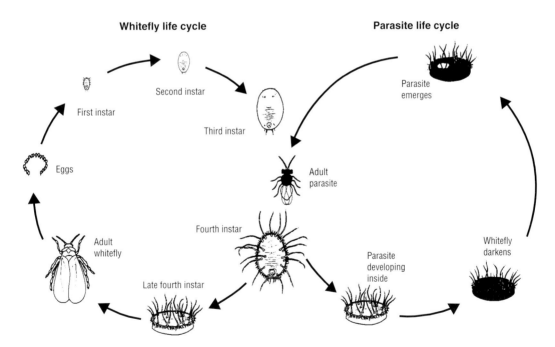

Whitefly life cycle   **Parasite life cycle**

Second instar

First instar

Third instar

Eggs

Adult parasite

Parasite emerges

Adult whitefly

Fourth instar

Whitefly darkens

Late fourth instar

Parasite developing inside

■ FIGURE 7-3. Life cycle of the greenhouse whitefly and *Encarsia formosa*. The parasite prefers to lay eggs in third-instar and early fourth-instar whiteflies, which darken (the black scale stage) about 7 to 10 days after becoming parasitized. The parasite develops through three larval stages within the whitefly nymph, then pupates, emerges as an adult, and flies off to parasitize other whiteflies.

These oblong brown to orangish pupae of *Leptomastix dactylopii* occur among mealybug wax. After emerging, the adult parasite leaves a round hole or hinged cap at the end of its pupal case.

LENGTH

The late-instar blue gum psyllids on this *Eucalyptus pulverulenta* leaf are parasitized by *Psyllaephagus pilosus*, causing them to appear bloated and mummified.

LENGTH

Pseudaphycus, and *Tetracnemoidea* (=*Arhopoideus*) spp., provide biological control of certain mealybug species (table 7-4). Longtailed mealybug is controlled in some situations by *Anarhopus sydneyensis*.

Citrus Mealybug Parasite

The citrus mealybug infests citrus and many greenhouse and indoor plants. The mealybug destroyer lady beetle and two parasitic wasps, *Leptomastix dactylopii* and *Leptomastidea abnormis* (Encyr-

tidae), can control citrus mealybug. The mealybug destroyer and *L. dactylopii* are commercially available.

Leptomastix dactylopii is a yellowish brown wasp that lays its eggs in late-instar nymphs and adult mealybugs. *Leptomastix* prefers hosts in warm, sunny, humid environments. It can complete one generation in 2 weeks at 85ºF (30ºC) or in 1 month at 70ºF (21ºC). Because *Leptomastix* is more effective at lower host densities than the mealybug destroyer lady beetle, these beneficials complement each other.

■

PSYLLID PARASITES

Psyllids (family Psyllidae), sometimes called jumping plantlice, are plant juice sucking Homoptera that look like miniature cicadas. With the notable exception of the pear psylla, most pest psyllids occur on native and introduced ornamental plants. Many species of psyllids are controlled by various lady beetles, lacewing larvae, small predaceous bugs, and parasitic wasps. For example, acacia psyllid is now controlled by an introduced minute pirate bug and lady beetle (table 7-4). The peppertree psyllid in California appears to be partially controlled by an introduced *Tamarixia* sp. (Eulophidae) wasp.

■

Blue Gum Psyllid Parasite

The blue gum psyllid damages *Eucalyptus pulverulenta* foliage grown commercially for floral arrangements and also infests young foliage on some eucalyptus species in landscapes. *Psyllaephagus pilosus* (Encyrtidae) introduced from Australia in 1993 has dramatically reduced blue gum psyllid populations, allowing most growers to eliminate insecticide applications on their eucalyptus crops (Dahlsten et al. 1996).

■

Eugenia Psyllid Parasite

The eugenia psyllid was accidentally introduced into California in 1988. Its feeding pits, distorts, and reddens

COMMON NAME	SCIENTIFIC NAME	NATURAL ENEMIES	REFERENCE
acacia psyllid	*Acizzia uncatoides*	*Anthocoris nemoralis* minute pirate bug *Diomus pumilio* lady beetle	Dreistadt and Hagen 1994
ash whitefly	*Siphoninus phillyreae*	*Clitostethus arcuatus* lady beetle *Encarsia inaron*	Dreistadt and Flint 1995; Gould, Bellows, and Paine 1992; Pickett et al. 1996
bayberry whitefly	*Parabemisia myricae*	*Eretmocerus, Encarsia* spp.	Rose and DeBach 1982, Walker and Aitken 1993
blue gum psyllid	*Ctenarytaina eucalypti*	*Psyllaephagus pilosus*	Dahlsten et al. 1996
citrophilus mealybug	*Pseudococcus fragilis*	*Coccophagus gurneyi, Hungariella pretiosa*	Compere and Smith 1932, DeBach 1964
citrus blackfly*	*Aleurocanthus woglumi*	*Amitus hesperidum, Encarsia opulenta*	Tsai and Steinberg 1991
citrus mealybug	*Planococcus citri*	*Leptomastix dactylopii, Leptomastidea abnormis* mealybug destroyer lady beetle	Bartlett and Lloyd 1958, Bennett et al. 1976
citrus whitefly	*Dialeurodes citri*	*Encarsia, Eretmocerus,* and *Prospaltella* (=*Encarsia*) spp.	Anonymous 1991a
Comstock mealybug	*Pseudococcus comstocki*	*Allotropa burrelli, A. convexifrons Pseudaphycus malinus*	Meyerdirk, Newell, and Warkentin 1981
eugenia psyllid	*Trioza eugeniae*	*Tamarixia* sp.	Dahlsten et al. 1995
longtailed mealybug	*Pseudococcus longispinus*	*Anarhopus sydneyensis*	Bartlett and Lloyd 1958, Flanders 1940
woolly whitefly	*Aleurothrixus floccosus*	*Amitus spiniferus, Cales noacki, Eretmocerus* sp.	Miklasiewicz and Walker 1990

Note: All natural enemy species listed here are parasitic wasps, except for the minute pirate bug and lady beetles. Consult the index for family names.

*Occurs in Florida and the Caribbean, but not in California.

foliage of Australian bush cherry or eugenia, a widely planted ornamental in coastal California. A *Tamarixia* sp. (Eulophidae) was introduced in a successful University of California classical biological control program partly funded by Disneyland, whose topiary hedges were being severely damaged by this pest (Dahlsten et al. 1995).

THRIPS PARASITES

Resident populations of predaceous thrips, minute pirate bugs, predaceous mites, and parasites help control plant-feeding thrips in natural situations (van Lenteren and Loomans 1995). Recently discovered larval parasites of western flower thrips include two tiny wasps (*Ceranisus americensis* and *Ceranisus menes*, Eulophidae) (Triapitsyn and

LENGTH

A *Tamarixia* sp. left an emergence hole in a dead peppertree psyllid shown here. Eugenia and California pepper tree each are infested with a different species of psyllids, which causes foliage to develop pits.

Headrick 1995) and a nematode (*Thripinema nicklewoodii*).

Greenhouse Thrips Parasite

Greenhouse thrips attack many agricultural, garden, and ornamental plants. To help control it, *Thripobius semiluteus* (Eulophidae) has been introduced into California from Australia and Brazil. *Thripobius* is now established in scattered locations along the central and southern coast of California, but apparently does not occur in northern California. Augmentative releases in greenhouse crops, avocados, and citrus are also employed by some growers (McMurtry, Johnson, and Newberger 1991). In Southern California, some growers make an inoculative release of several hundred *Thripobius* per 15-foot-tall avocado tree when young thrips are first observed.

Thripobius lays one egg in each first-instar or early second-instar thrips nymph. Parasitized thrips die as second instars, and the parasite pupae remain on the surface where thrips were feeding. Parasitized thrips turn black, in contrast to the yellow unparasitized greenhouse thrips nymphs. Because *Thripobius* develops from egg to adult in about 3 weeks when temperatures average 70°F (21°C), many generations can occur each year in warm areas.

LEAFMINER PARASITES

Parasitic wasps often control *Liriomyza* spp. leafminers in natural habitats. Parasites can be abundant in various ornamentals and later in the crop production cycle when they migrate into cole crops and lettuce. Leafminer outbreaks frequently occur after insecticides applied to kill other pests kill the parasites of leafminers.

A black and yellow adult *Thripobius* wasp among parasitized (black) and unparasitized (yellow) greenhouse thrips nymphs. Parasitized thrips blacken and swell around the head as the wasp larva matures inside. Adult greenhouse thrips are also black, but adult thrips are mobile and larger than the parasitized nymphs, which don't move.

An adult female of *Diglyphus begini* seeks a leafminer larva by puncturing a leaf with its ovipositor. *Diglyphus* spp. adults host feed on second-instar leafminers and lay eggs near third instars, in both cases causing the leafminer to die. Although adults are small, if parasites are abundant, adults can sometimes be observed flying in groups that hover near infested plants. Adults also are caught in yellow sticky traps used to monitor adults of leafminers and certain other pests and their parasites.

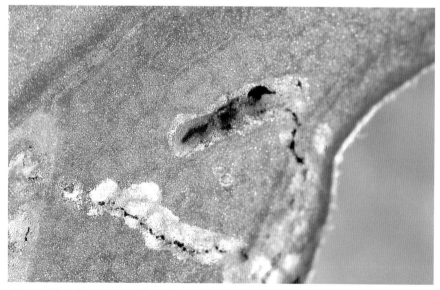

A greenish *D. begini* parasite larva feeding on an orangish leafminer larva viewed here through the leaf surface.

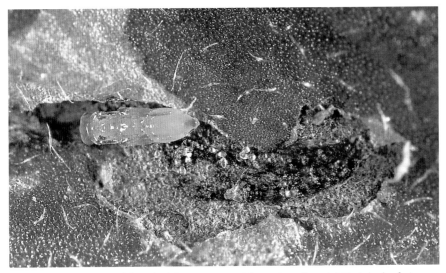

A greenish *D. begini* pupa exposed in a mine where it killed the leafminer larva. Pupae of leafminer parasites usually have a relatively flat body with a large head and large eyes and are naked within the leaf mine. Leafminers pupate within their last larval skin and often pupate outside of the mine on the leaf or soil surface.

Certain leafminer parasites (such as *Diglyphus begini*, *Diglyphus isaea*, and *Dacnusa sibirica*) may be commercially available. Their release in greenhouses has controlled leafminers infesting marigolds grown for seed (Heinz, Newman, and Parrella 1988; Heinz and Parrella 1990), cut flower chrysanthemums (Heinz, Nunney, and Parrella 1993; Jones, Parrella, and Hodel 1986), and certain vegetable crops.

PARASITIC FLIES

Flies (order Diptera) are second only to the Hymenoptera in their importance as parasites. An estimated 16,000 species of flies are parasites, about 20% of all known parasitic insects (Feener and Brown 1997). At least 12 families of flies contain parasitic species (table 7-5).

Unlike parasitic wasps, most species of flies do not have a well-developed egg-laying organ (ovipositor) that can insert eggs in or onto their hosts. Many parasitic flies lay their eggs or larvae on plants. Some species then enter their hosts by being eaten. For example, a caterpillar consumes plant tissue containing parasitic fly eggs, which hatch and develop inside the host's gut. In many species, eggs hatch into mobile, free-living larvae called planidia that seek out hosts. These first-instar larvae are often dark colored, flattened, and have well-developed spines or mouthparts. After finding their hosts, planidia penetrate them and feed inside as endoparasites. In other species, planidia attach themselves to the outside, feeding externally by inserting only their mouthparts into the hosts. Once in or on their hosts, planidia change greatly in appearance. Later-stage parasitic fly larvae are often pale colored and grublike, without conspicuous spines or mouthparts.

The biology and appearance of parasitic flies varies greatly among the thousands of known species. The family Tachinidae, the most important family of flies providing biological control (Arnaud 1978), are emphasized here. Unlike most other parasitic flies, many tachinids have specialized

This colorful *Trichopoda pennipes* lays its oval pale eggs on squash bugs, commonly on the bug's side. This tachinid parasite has been introduced into California from the northeastern U.S. by the California Department of Food and Agriculture's Biological Control Program (Pickett, Schoenig, and Hoffmann 1996; Schoenig and Pickett 1996).

Tachinid pupae are commonly oblong and dark reddish. These pupae of *Uramyia halisidotae* emerged from silverspotted tiger moth larvae infesting Monterey pine. The round grayish material is caterpillar frass (fecal droppings) covered with decay fungi.

TABLE 7-5. Selected fly families with parasitic larvae.

Acroceridae, small-headed flies

Internal parasites of spiders. Adults appear hunchbacked, with a small round head and globular abdomen. Eggs are laid on substrates in small clusters. Larvae hatch, move, and enter any spiders they contact. At least 250 species are known.

A

Bombyliidae, bee flies

Mostly internal and external parasites of Lepidoptera and Hymenoptera larvae; some species attack larvae of beetles, flies, and moths or eggs of grasshoppers. Most adults are stout, densely hairy, medium to large flies with a long, thin mouthpart (proboscis). Eggs are laid near hosts. Larvae crawl to hosts and enter them. About 3,000 species are known.

B

Chironomidae, midges

Most species have aquatic larvae that feed on decaying organic matter; others are external parasites of aquatic invertebrates, including mayflies and snails. Adults look like small, delicate mosquitoes and often swarm near water.

C

Conopidae, thick-headed flies

All are internal parasites, mostly of adult bees and wasps. Adults often pursue their host, grab it in flight, and insert their egg into the adult host's abdomen. Adult flies are medium sized, often yellow or brown, with a long slender abdomen and a head broader than their thorax. About 500 species are known.

D

Cryptochaetidae, cryptochaetid flies

All are internal parasites of scale insects. Only one species occurs in North America, *Cryptochaetum iceryae*, a small blackish blue fly that is an important parasite of cottony cushion scale.

E

Nemestrinidae, tangle-veined flies

Most species are internal parasites of locusts and beetle larvae and pupae. Adults are medium sized and stout-bodied, with long mouthparts. At least 250 species are known.

F

Phoridae, humpbacked flies

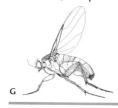

Most species eat decaying organic matter; some are internal parasites of ants, bees, caterpillars, crickets, termites, moth pupae, and fly larvae. Adults are small to minute, with an enlarged, humped thorax and few wing veins.

G

Pipunculidae, big-headed flies

All are internal parasites, primarily of leafhoppers and planthoppers. Adults are small with a very large head composed mostly of eyes. About 400 known species.

H

Pyrgotidae, pyrgotid flies

Internal parasites of adult June beetles and other Scarabaeidae. Adults are large, elongated flies with banded or colored wings; like their hosts, they are nocturnal and are rarely seen. About 5 species are known in the United States, all in the East.

I

Tachinidae, tachinid flies

Most are internal parasites of immature beetles, butterflies, and moths. Other hosts include earwigs, grasshoppers, and true bugs. Adults are often dark, robust, hairy flies resembling a house fly, but with very stout bristles at the tip of their abdomen. Eggs are laid on hosts or on plants. Eggs on plants are eaten and then hatch inside the host or larvae hatch on plants and enter hosts that approach. Over 1,500 species are known.

J

Sources: Borror, De Long, and Triplehorn 1981; Cole 1969; Daly, Doyen, and Ehrlich 1978 (A–D, F–J *from* Cole 1969, C by Jones, D by Celeste Green, I by C. S. Papp, reprinted with permission from University of California Press; E *from* Essig 1926, by S. W. Williston).

abdominal structures for laying eggs in or on their hosts.

CATERPILLAR PARASITES

Most species of moth and butterfly larvae (caterpillars) are attacked by one or more species of parasitic wasps or flies.

Noctuid and Pyralid Parasites

At least a dozen species of Noctuidae and Pyralidae moths throughout North America and Europe are attacked by the tachinid *Voria ruralis*. Cabbage looper is a preferred host, but the alfalfa looper, fall armyworm, and variegated cutworm are also parasitized. Other similar parasites also attack these pests; for example, at least 19 species of Tachinidae in North America attack *Trichoplusia* spp. loopers.

Eggs of *V. ruralis* hatch within several minutes of being laid just beneath the caterpillar's cuticle or outer "skin." A caterpillar may be "stung" several times as the parasite lays more than one egg. After killing its host, the larvae emerge, drop to the ground, and form oblong pupae, which are dark red and ⅓ inch (8 mm) long. One parasite generation requires about 10 to 30 days, depending on the temperature.

Leafroller Parasite

Erynnia (=*Anachaetopsis*) *tortricis* (Tachinidae) parasitizes larvae and pupae of over two dozen different caterpillars in at least seven different families, mostly small moths in the family Tortricidae. Hosts of *E. tortricis* include amorbia, codling moth, obliquebanded leafroller, omnivorous leafroller, oriental fruit moth, peach twig borer, pink bollworm, and sunflower moth. The parasite adult is shiny, dark grayish, and about ¼ inch (6 mm) long. This endoparasite lays one to several eggs on the head or thorax of host larvae. Parasite larvae do not kill the host until after the host pupates. Parasitized moth pupae may then be recognized by the

1⅜ INCHES
LENGTH

The dark blotches on this cabbage looper are tachinid parasite oviposition stings. Photo by Earl R. Oatman.

LENGTH

Adults of many beneficial tachinid species are robust, dark, and look like house flies, except that tachinids have stout bristles at the tips of their abdomens. This *Erynniopsis antennata* parasite is approaching an elm leaf beetle larva.

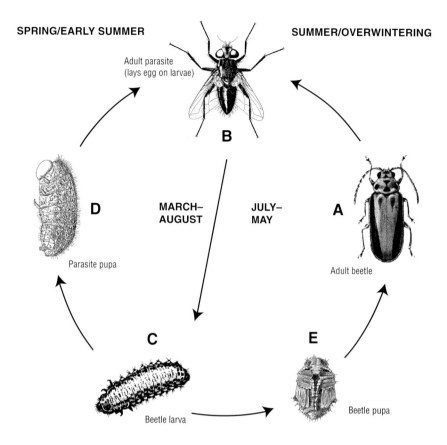

SPRING/EARLY SUMMER

SUMMER/OVERWINTERING

Adult parasite
(lays egg on larvae)

B

D

Parasite pupa

MARCH–
AUGUST

JULY–
MAY

A

Adult beetle

C

Beetle larva

E

Beetle pupa

■ FIGURE 7-4. The tachinid parasite *Erynniopsis antennata* develops through one of two different phases, depending on the season.

SPRING/EARLY SUMMER

A: The parasite spends the winter as first-instar larvae inside overwintering adult elm leaf beetles. B: *Erynniopsis* complete development after beetles begin feeding on foliage in the spring and adult parasites emerge from adult beetles; this parasite emergence is not readily observed. C: Each fly lays one egg on each of several dozen beetle larvae. Eggs hatch and the parasite larvae enter their host and feed inside. D: During spring and early summer parasitized beetle larvae are killed. *Erynniopsis antennata* then develop into black to reddish, cylinder- or teardrop-shaped parasite pupae at the tree base among yellowish beetle pupae.

SUMMER/OVERWINTERING

E: As the season progresses, an increasing proportion of the parasite population enters diapause (their inactive, overwintering phase), remaining as immature parasites inside beetle pupae. A: Beetles undergo metamorphosis into overwintering adults. B: Parasites mature and emerge from beetles in spring and this cycle repeats. One parasite generation takes several weeks during summer and 6 months or more over winter. Several generations occur each year.

Sources: Dreistadt and Dahlsten 1990b, Luck and Scriven 1976 (A *from* Anonymous 1960; B *from* Cole 1969, reprinted with permission from University of California Press; C *from* Anonymous 1979; D *from* Silvestri 1910; E *from* Herrick 1913).

prominent Y-shaped parasite spiracles that protrude beneath the tips of the wing pads of the host pupa.

BEETLE PARASITES

Immature stages (egg, larva, and pupa) of most beetles are attacked by parasites. Parasites of alfalfa weevil and lady beetles are illustrated in the discussion of parasitic wasps earlier in this chapter. Two fly parasites of beetles are discussed here.

Elm Leaf Beetle Parasites

Erynniopsis antennata (Tachinidae), which occurs throughout the United States, is the most important parasite of elm leaf beetle in California. Female flies deposit one egg on a late-instar beetle larva. The maggot enters the host and feeds and pupates inside (fig. 7-4). These tachinid pupae are sometimes attacked and killed by the gregarious (several per host) secondary parasite *Oomyzus* (=*Tetrastichus*) *erynniae* (Eulophidae). Two other beneficial *Oomyzus* (=*Tetrastichus*) spp. parasitize elm leaf beetle pupae (*O. brevistigma*) and eggs (*O. gallerucae*). Application of the bacterium *Bacillus thuringiensis* subspecies *tenebrionis* (=*san diego*) (see chapter 9) and spot application or bark banding with insecticide (chapter 1 and table 1-3) are control actions for elm leaf beetle that are compatible with the use of natural enemies (Dahlsten et al. 1993).

Japanese Beetle Parasites

The tachinid *Hyperecteina aldrichi* is the most easily recognized species of at least five different parasites introduced to attack Japanese beetle in the eastern United States. Parasites are of limited effectiveness in controlling the Japanese beetle. However, *Bacillus* spp. and entomopathogenic nematodes discussed in chapter 9 selectively kill Japanese beetle larvae and are compatible with this parasite of adult beetles.

In the western United States, report suspected Japanese beetles to the county department of agriculture. Officials will conduct eradication to prevent this pest from becoming established.

■ COTTONY CUSHION SCALE PARASITES

A parasitic fly (*Cryptochaetum iceryae*, family *Cryptochaetidae*), and the scale-feeding vedalia beetle discussed in chapter 8, were imported from Australia into California in the late 1880s to control cottony cushion scale. They now provide complete biological control of cottony cushion scale on numerous hosts in California and elsewhere in the world, unless disrupted by adverse conditions such as pesticide applications. *Cryptochaetum* apparently predominates in coastal areas; vedalia is most abundant in desert citrus-growing regions (Quezada and DeBach 1973). Either or both species can occur in interior areas of California if cottony cushion scale is present.

The tiny adult cryptochaetid is a dark blue or green to black fly with short, rounded, grayish wings. One fly generation requires about 1 month outdoors in summer, with up to five or six generations developing per year.

EGGS

Many species of tachinids lay oblong white eggs on top of the head or thorax of their host. Two eggs of *Hyperecteina aldrichi* are visible on the thorax of one of these Japanese beetles. The parasite maggot bores downward from the egg into the host's body. This species can mature and kill its host within a week, preventing parasitized beetles from laying eggs.

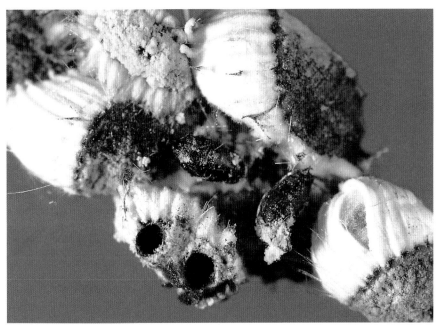

LENGTH

Pupae of the cottony cushion scale parasite *Cryptochaetum iceryae* may appear outside dead scales or remain inside scales, leaving one or more round emergence holes. Two externally pupating, oblong black parasite pupae and emergence holes from two internally pupating parasites are visible here. The female *Cryptochaetum* parasite lays one egg in small scales and a dozen or more in larger hosts, which provide more food. The larva feeds inside, eventually forming an orangish to black pupa with two tiny protruding breathing tubes (spiracles).

CHAPTER EIGHT

PREDATORS OF ARTHROPODS

Magnifying glass signifies that subject is less than 1 mm.

PREDATORS FEED ON MORE than one individual host during their lifetimes and are usually larger or stronger than the animals they eat. Many predators feed on a variety of insects, mites, and other invertebrates. Although these "general predators" reduce pest populations and help prevent pest outbreaks, by themselves they generally do not provide complete biological control. Parasites and other more specialized predators, such as those feeding mainly on a group of closely related pests, are often especially effective at controlling specific pests. Some predaceous invertebrates also feed on pollen, nectar, and honeydew in addition to prey species.

RECOGNIZING PREDATION

Because they are relatively large and generally do not live within their hosts, most predators are more readily observed than parasites. However, once predators have left the site, evidence of predation may not be conspicuous because pests can die from other causes, and some predators remove or entirely consume their prey, leaving no evidence of predation.

In addition to active predator adult and larval stages, look for egg and pupal stages and cast skins of predators near their prey. Although chewing predators may entirely consume their host, empty, collapsed host skins are often left behind by predators with sucking mouthparts. If recent plant damage, hatched pest eggs, or old cast skins of younger-stage pests are visible, but few or no older pest stages are present, the pests may have been killed by predators.

Some predators disguise their appearance to protect themselves from their own natural enemies. For instance, larvae of some lady beetles that eat mealybugs or woolly aphids exude over themselves flocculent wax, which makes them look like their prey and may protect them from predators such as ants. Burgundy-colored adult *Chilocorus bipustulatus* lady beetles look like the dome-shaped lecanium scales they commonly feed on; like many scale predators, their larvae feed hidden under the scale's body. The numbers of other predators, such as predaceous ground beetles and some nonweb-making spiders, may be underestimated because these predators hide during the day and hunt their prey mostly at night.

BEETLES

Beetles, order Coleoptera, are the most diverse of all insects, with over a quarter of a million known species in about 150 families (Papp 1984). About 40 beetle families contain predaceous species; common predatory groups are listed in table 8-1.

Lady beetles (Coccinellidae) and predaceous ground beetles (Carabidae) are probably the two most important groups of predatory beetles. Some beetles (such as many carabids) are relatively general feeders and will prey on virtually any invertebrate they encounter and can capture. Certain beetles are more host-specific: for example, many species of lady beetles feed mostly on aphids. Certain other beetles are very host specific, such as the vedalia lady beetle, which preys only on cottony cushion scale.

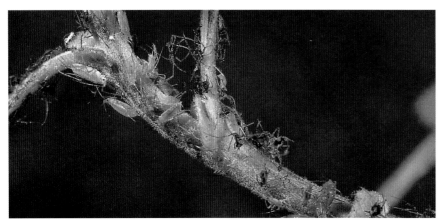

These collapsed, greenish aphids have recently been killed by having their body contents sucked out. The shriveled black aphids are older evidence of predation. The orange predaceous midge larvae (*Aphidoletes* sp.) visible here caused this predation, but other predators, including lady beetles and lacewing larvae, can leave similar predation evidence.

LENGTH ⊢⊣

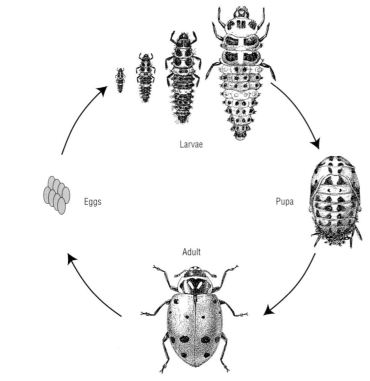

Larvae

Eggs

Pupa

Adult

■ FIGURE 8-1. Life cycle of the convergent lady beetle. Development from egg to adult takes about 3 to 6 weeks, with reproduction arrested during overwintering. Adapted from illustrations by F. H. Chittenden in Sanderson and Jackson (1912).

LADY BEETLES (LADYBUGS)

Lady beetles (Coccinellidae), commonly called "ladybugs," are easily recognized by their shiny, convex, half-dome shape. Lady beetles have short, clubbed antennae; some species are brightly colored while others are dark with few or no markings.

About 500 species of lady beetles occur in America north of Mexico (Gordon 1985). Most are predaceous as both adults and larvae (Hagen 1962, 1974). Two notable exceptions are the Mexican bean beetle and the squash beetle, pests of beans and cucurbits, respectively, in the eastern and midwestern United States. No plant-feeding lady beetles are known pests in California. A few lady beetles feed on fungi, but most eat mites or soft-bodied insects. When prey are scarce, most adults can survive (although generally not reproduce) on nectar, honeydew, pollen, or a combination of these.

Many lady beetles have a life cycle similar to that of the convergent lady beetle (fig. 8-1), except that their prey varies. In addition, some populations of the convergent lady beetle migrate to overwintering sites (fig. 8-2), but populations of most lady beetle species do not migrate and form overwintering aggregations. Most species overwinter as adults in protected places on or near host plants. Some are active year-round in mild winter areas, although their reproduction and abundance changes seasonally in association with that of their prey. Many species of lady beetles specialize on certain types of insects or mites, making them especially effective predators. Young lady beetle larvae usually pierce and suck the contents from

TABLE 8-1. Common predatory beetle families.

Cantharidae, soldier beetles or leather-winged beetles

Larvae are predaceous. Adults of some species feed on pollen and nectar and are often found on flowers; others are predaceous on aphids and insect eggs and larvae or feed on both flowers and insects. Adults are elongate, often with a red, orange or yellow head and abdomen and black, gray, or brown soft wing covers. Larvae are dark, elongate, and flattened. They feed under bark or in soil or litter, primarily on eggs and larvae of beetles, butterflies, moths, and other insects. Over 100 species in California.

Carabidae, predaceous ground beetles

Most are predaceous as adults and larvae; a few species are parasitic. Adults are primarily nocturnal, have long legs, and are usually dark, although some are colored. Larvae often occur in litter or soil, are elongate, usually have 10 well-defined segments, and taper toward their rear ends. Over 40,000 known species in the world.

Cicindelidae, tiger beetles

Adults stalk their prey, have long legs, and run rapidly or fly quickly when disturbed. They prefer open, sandy, sunny locations. Adults' wing covers are often patterned and brightly colored. Larvae have strong spines and large jaws. Larvae usually wait in a vertical tunnel in the soil and snatch passing insect prey. About 40 species in California.

Cleridae, checkered beetles

Adults and larvae of most species are predaceous; a few are scavengers or parasites. Adults are often brightly colored and hairy; bark beetles and other wood-boring insects are common prey. Eggs are usually laid under bark, where larvae feed on immature bark beetles. Larvae are elongate, flattened, often have numerous small spines on all segments, and usually have distinct pigment that is especially dark around the head. About 100 species in California.

Coccinellidae, lady beetles or "ladybugs"

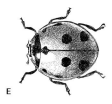

Most adults and larvae feed on soft-bodied insects or mites. Adults are dome-shaped, ranging from tiny to medium-sized and are from black to brightly colored. Larvae are active, elongate, have long legs, and resemble tiny alligators. About 500 species in the United States and Canada.

Dytiscidae, predaceous diving beetles

Larvae are aquatic and adults occur in or near water. Both stages are predaceous on aquatic insects, small fish, and tadpoles. Adults are smooth, oval, somewhat flattened, and dark green, black, or brown. Larvae are elongate, widest around the thorax, have long jaws, and are very active. About 125 species in California.

Gyrinidae, whirligig beetles

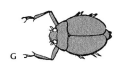

Adults primarily scavenge insects that fall on the water surface. Larvae are predaceous on insects and other small aquatic animals. Larvae look like small centipedes: elongate, flattened, and pale except for their dark head and thorax. Adults are oval black beetles that often swim in circles and occur in groups. About a dozen species in California.

Melyridae, soft-winged flower beetles

Most adults and larvae are predaceous, feeding on aphids, leafhoppers, and immature stages of insects in field crops, including alfalfa, corn, and cotton. Adults are elongate to oval soft-bodied beetles, often brightly colored red or brown and black; often they appear velvety due to many fine hairs. The soil-dwelling larvae are elongate, flattened, and have a large head and distinct mandibles. About 450 species in North America.

Staphylinidae, rove beetles

Most are predaceous and live in litter as adults and larvae. Larvae of some species are parasitic. Prey includes insect eggs, larvae, pupae, and small, soft-bodied species such as aphids and mites. Adults are usually brown or black and have short wing covers that expose abdominal segments visible from above at their rear. Larvae are long and thin with a prominent head. About 3,100 species in North America.

Sources: Borror, De Long, and Triplehorn 1981; Papp 1984 (A, F *from* Daly, Doyen, and Ehrlich 1978, reprinted with permission of H. V. Daly; B–D, G from Packard 1876; E *from* Sanderson and Jackson 1912, by F. H. Chittenden; H *from* Wildermuth 1914; I *from* Quayle 1932).

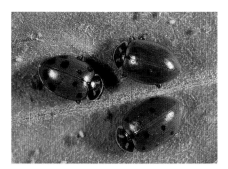

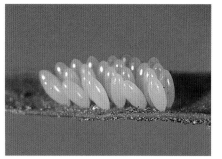

Adult convergent lady beetles, *Hippodamia convergens*, are mostly orange and often have 13 black spots; however, many individuals have fewer spots and some have none. A similar widespread species, *Hippodamia quinquesignata* (not shown), is always spotless and may or may not have converging white marks on its thorax.

Lady beetle eggs, like these of *H. convergens*, are usually oblong or spindle-shaped. Depending on the species, these football-shaped eggs are laid individually or in compact clusters of up to several dozen on pest-infested plants. Similar-looking leaf beetle eggs are distinctly wider at the base where they are attached to the leaf.

The parenthesis lady beetle (*Hippodamia parenthesis*) is a common aphid feeder throughout the United States. Its head and thorax are black, except for indented white markings along the thoracic margins. The wing covers are orangish with black markings; the hindmost pair of markings may look like a pair of parentheses (as shown here) or be broken black blotches.

The elongate, blackish, orange-spotted, alligator-shaped convergent lady beetle larvae develop through four instars, each lasting about 3 to 7 days. Also visible here among whitish aphid cast skins are the immobile, oblong, orangish to dark beetle pupae. Adults emerging from pupae feed for 1 or 2 weeks before laying the first of several hundred eggs.

their prey. Older larvae and adults chew and can consume the entire prey.

APHID-FEEDING LADY BEETLES

Most of the larger, reddish orange lady beetles feed primarily on aphids. Some lady beetles that are not red also eat aphids. Several species of *Hippodamia* and *Coccinella*, *Adalia bipunctata*, and *Olla abdominalis* (=*O. v-nigrum*) are probably the most common aphid- feeding lady beetles in the western United States.

Recently introduced species, including *Harmonia axyridis* and *Coccinella septempunctata*, are becoming very important predators. *Coleomegilla maculata* (Giroux, Duchesne, and Coderre 1995) and certain other lady beetles important in the eastern United States are uncommon or do not occur in the West.

Adults of many species of lady beetles can consume about 100 aphids per day (Hodek 1973). Actual consumption varies greatly, depending on numerous factors including temperature, host

plant, season, physiological condition of the beetles, and the size, species, life stages, and density of the aphids and beetles (Hagen 1962, 1974).

Convergent Lady Beetle

The familiar convergent lady beetle (*Hippodamia convergens*) can be common wherever aphids are abundant, including gardens, landscapes, trees, and field crops. It feeds primarily on aphids, but also to some extent consumes whiteflies (Hagler and Naranjo 1994), other soft-bodied insects, and insect eggs. About two dozen *Hippodamia* spp. occur in the United States; they are among the most common orangish and black beetles that eat aphids.

Many convergent lady beetles in the western United States overwinter in large aggregations in the Sierra Nevada and elsewhere. In early spring after temperatures warm, adults migrate with prevailing winds from the mountains to coastal areas or the Central Valley of California. If aphids are abundant, several generations of beetles may develop. When aphids become scarce, many adults migrate to the mountains and form overwintering aggregations.

This aggregation habit allows convergent lady beetles to be readily collected and sold for release to control aphids. But, because of their innate tendency to

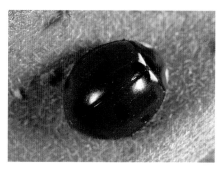

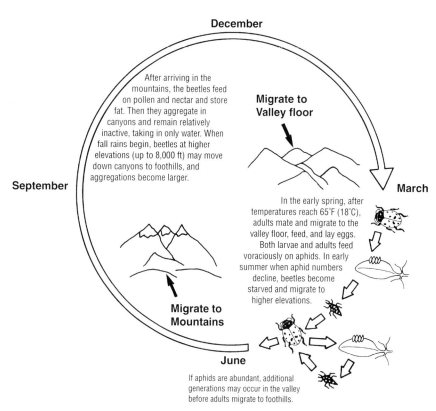

After arriving in the mountains, the beetles feed on pollen and nectar and store fat. Then they aggregate in canyons and remain relatively inactive, taking in only water. When fall rains begin, beetles at higher elevations (up to 8,000 ft) may move down canyons to foothills, and aggregations become larger.

December

September

Migrate to Valley floor

March

In the early spring, after temperatures reach 65°F (18°C), adults mate and migrate to the valley floor, feed, and lay eggs. Both larvae and adults feed voraciously on aphids. In early summer when aphid numbers decline, beetles become starved and migrate to higher elevations.

Migrate to Mountains

June

If aphids are abundant, additional generations may occur in the valley before adults migrate to foothills.

LENGTH |—|

Coccinella species are a major group of aphid-feeding lady beetles. About 12 species of *Coccinella* occur in the United States (Gordon and Vandenberg 1995). Like most non-*Hippodamia* species of lady beetles, *Coccinella* have a more rounded shape and their legs are less obvious because the tips of their femoral leg segments are hidden when viewed from above (fig. 8-3). This sevenspotted lady beetle (*C. septempunctata*) was introduced from Europe and India (Maredia et al. 1992, Phoofolo and Obrycki 1995). Its thorax is black with white along the front margin. There are seven large black spots on its red or orangish wing covers, which may have some white near the front.

■ FIGURE 8-2. The annual cycle of convergent lady beetle migration and aggregation in California allows collectors to harvest large numbers from the Sierra Nevada for sale.

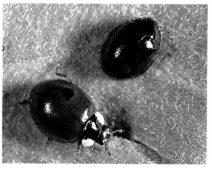

LENGTH |—|

This ninespotted lady beetle (*Coccinella novemnotata*) occurs throughout the United States in forms with nine large spots, a mixture of large and small spots, or no spots. This nonspotted form is the most common in the West and is easily confused with the California lady beetle (*Coccinella californica*), which occurs only in Pacific Coast states and British Columbia. Wing covers of these aphid-feeding *Coccinella* spp. are almost entirely reddish, sometimes with a dark spot or narrow band along the front where the wing covers meet. The head and thorax of both species are black with a squarish white blotch along each front outer margin of their thorax. Unlike the California lady beetle, the ninespotted lady beetle has a pale area between its eyes and along the front of the pronotum connecting the white marginal thoracic spots.

LENGTH |—|

Leaf beetles (family Chrysomelidae) such as this striped cucumber beetle (below) are common pests that can be mistaken for the similarly shaped lady beetles such as *Hippodamia sinuata* (top). Chrysomelids can be distinguished by their tarsi or feet segments (fig. 8-3) and their longer antennae. Lady beetles have short antennae, which usually have a clubbed end.

LENGTH |—|

The twospotted lady beetle (*Adalia bipunctata*) feeds on aphids and is common in trees and gardens throughout North America and Europe. Two forms of this variable beetle are shown. One form is black with four red or orange spots on the wing covers and sometimes orange margins on the thorax. The other form has mostly orange or red wing covers with two black spots; its thorax is black with two white marginal blotches.

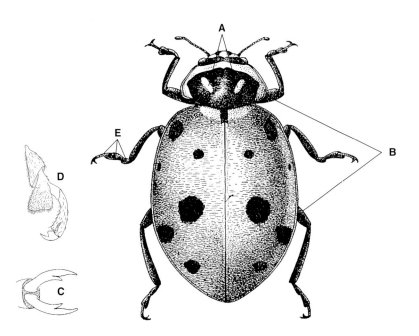

The ashy gray lady beetle (*Olla v nigrum −O. abdominalis*) is a common arboreal aphid-feeder throughout the United States. It also occurs in vegetables and fields. Its thorax and wing covers are gray to pale yellowish with black spots. A black form with two large red spots also occurs.

LENGTH

■ FIGURE 8-3. How to recognize *Hippodamia* spp. and distinguish lady beetles from leaf beetles: A: The convergent lady beetle (*Hippodamia convergens*) is named for the two converging white marks on its thorax, although some forms of at least six other *Hippodamia* spp. and some other lady beetles have similar marks. B: *Hippodamia* spp. appear more leggy than other lady beetles because the tips of the femur portion of their legs clearly extend beyond their body when viewed from above. C: *Hippodamia* spp. also can be distinguished from similar-looking lady beetles because their tarsal claws are cleft or forked. D: The feet of lady beetles in other genera have a single narrow to broad claw or pad. E: Lady beetles have three distinct tarsal segments on their feet. Leaf beetles (Chrysomelidae) appear to have four tarsal segments on their feet. Because leaf beetles have a body shape that resembles lady beetles, examining their feet helps to distinguish lady beetles from leaf beetles, which feed on plants. Adapted from Hagen 1982, by Celeste Green (C, D *from* Gordon 1985, reprinted with permission from the American Museum of Natural History). See fig. 8-6 for an illustration of how to distinguish between male and female convergent lady beetles.

The ashy gray lady beetle larva has yellowish spots on its thorax and abdomen that vary in size and location according to instar.

LENGTH

LENGTH

The multicolored Asian lady beetle (*Harmonia axyridis*) is a predator of aphids and scales that has been introduced from Asia (Coderre, Lucas, and Gagné 1995; LaMana and Miller 1996). It has more than 100 forms with different spots and colors. Shown here are two common orangish forms with 19 dark spots and a form without spots.

LENGTH

A pupa, late-instar larva, and reddish-form adult multicolored Asian lady beetle. During their lifetime, each female lays several dozen to several hundred eggs; each lady beetle adult and larva can eat several dozen to several hundred aphids. Egg laying and feeding are highest when aphids are most abundant.

disperse, most lady beetles will soon leave the site where they are released, even if aphids are abundant. Releasing sufficient numbers of lady beetles can temporarily reduce aphid numbers in greenhouses and on small plants in localized areas (Dreistadt and Flint 1996, Raupp et al. 1994), but large-scale releases in the field have not been shown to be effective. There is no research-based information on how many beetles to release or how often to release them for most situations. Purchased beetles can be stored in the refrigerator (don't freeze them!), periodically warmed and allowed to drink by misting them with water, then released on small plants as needed.

Wetting plants first and releasing beetles on the ground under plant stems or trunks in the evening when it is cooler may slow beetle dispersal.

SCALE-FEEDING LADY BEETLES

Scale insects are preyed upon by a great variety of lady beetles, including *Chilocorus, Hyperaspis,* and *Rhyzobius* spp. The larvae of many scale-feeding species are easily overlooked because they often feed hidden underneath the scale body or cover (fig. 8-4). Adults are about 1/16 to 1/5 inches (1.5–5 mm) long and range from colorful to plain and dark.

■

Vedalia

The vedalia beetle (*Rodolia cardinalis*) feeds on only one species, the cottony cushion scale. Vedalia and the parasitic fly *Cryptochaetum iceryae* were introduced from Australia during 1888 and 1889. They saved California's fledgling citrus industry from being destroyed by the prolific cottony cushion scale. This first great success of classical biological control was later repeated in many other countries (fig. 8-5). Vedalia's success spawned subsequent projects of introducing natural enemies against other exotic pests throughout the world.

■

Black Lady Beetle

The black lady beetle (*Rhyzobius forestieri* =*Rhizobius ventralis*) and a similar-looking, smaller species (*Rhyzobius lophanthae* =*Lindorus lophanthae*) were introduced from Australia into the United States in the late 1800s. *Rhyzobius lophanthae* now occurs throughout the southern United States; the black lady beetle is reported only in California.

Both species are important predators of scale insects and mealybugs. Because younger larvae commonly feed on eggs and crawlers underneath the female scale's cover or body (fig. 8-4), young larvae may be difficult to see. Older larvae and adults feed openly on most scale stages. Predator preference for

LENGTH

Two species of scale-feeding lady beetles and one of their hosts, mature European fruit lecanium scales. The black lady beetle adult (bottom left) and its larva (center) are black to dark brown. The adult *Chilocorus bipustulatus* (bottom center) is red or brown with three yellowish bands or spots on each wing cover; its spiny larva (bottom right) is brownish tan and black (Hattingh and Samways 1994).

LENGTH

Black predominates on some vedalia, while others have more red. Their covering of fine hairs often gives them a grayish appearance unless viewed closely with good light or under magnification. Females lay oblong, red eggs singly or in groups on or near cottony cushion scales, such as the mature female scale shown here. The young reddish larvae emerge and feed on scale eggs and crawlers; more mature larvae and adults feed on all scale stages.

IDENTIFYING LADY BEETLES

Many species of lady beetles and other insects can be reliably identified only by using certain characteristics not readily distinguished by nonspecialists. Lady beetles are correctly identified by examining the complex and uniquely shaped male genitals as illustrated in Gordon (1985). Reproductive appendages are also used to distinguish between male and female beetles. Because of their often familiar colors, we describe species here based on color or spots. We also show how color differences can be used to distinguish the sex of convergent lady beetles and mealybug destroyers (fig. 8-6). However, color and spots vary and cannot be relied on for identification.

For example, four similar-looking species discussed in this book (*Axion plagiatum, Chilocorus orbus, C. kuwanae,* and one form of *Olla abdominalis* or *O. v-nigrum*) are shiny black with two red spots on their wing covers. Although the characteristics in Gordon (1985) and Gordon and Vandenberg (1991) are the only reliable method of distinguishing these species, *O. v-nigrum* has white margins on its thorax while the thorax of the other three species is entirely black. The location and size of red spots helps distinguish among the other three species. Spots of *C. orbus* are located in the front one-third of the elytra (wing covers). Spots of *A. plagiatum* are further forward, almost reaching the front margin of the elytra, and are larger than the spots on *C. orbus*. On *C. kuwanae*, spots are at or slightly behind the middle of the elytra.

Although it may sometimes be possible to distinguish even similar-looking lady beetles based on subtle differences in appearance, relying on color and spots to identify insects can fool even the experts. The multicolored Asian lady beetle (*Harmonia axyridis*) has more than 100 forms with different spots and colors. Historically, dozens of these forms were given different names by various entomologists who failed to recognize that they all were one species.

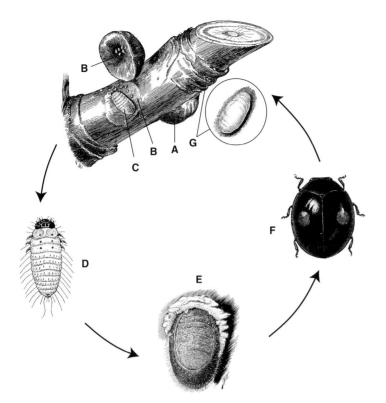

scale stage and species at least partly depends on scale cover hardness. Adult black lady beetles feed on all stages of oleander scale and apparently provide good control of this pest. The black lady beetle is less effective against California red scale because these scales spend much of their lives as females with hard covers that are resistant to beetle attack (Honda and Luck 1995).

Both *Rhyzobius* spp. have black to dark reddish brown wing covers. *Rhyzobius lophanthae* is about ¹⁄₁₆ to ¹⁄₈

This *Axion plagiatum* adult and larva are eating phylloxera (aphidlike insects) infesting a valley oak. Adults of *Chilocorus orbus, C. kuwanae,* and one form of *Olla v-nigrum* resemble this species; all are shiny black with two red spots. Older larvae of *A. plagiatum, C. orbus,* and *C. kuwanae* also resemble each other and are mostly dark with a pale head and abdominal area.

■ FIGURE 8-4. Many scales are preyed upon by lady beetle larvae that are easily overlooked because they feed hidden underneath scales: A: The female lecanium scale is a dome-shaped, immobile insect. B: Scale eggs and nymphs occur underneath and are revealed by lifting the mature female scale from bark. C: Young beetle larvae feed on immature scales beneath the female. Beetle larvae develop through four instars. D: Late-instar larvae feed on both immature and mature scales, and mature beetle larvae can occur openly on bark. E: Beetles pupate on or near the host plant. F: This adult *Hyperaspis binotata* is one of over 100 *Hyperaspis* spp. that occur in America north of Mexico. These mostly small dark beetles with lighter markings feed on various Homoptera, especially mealybugs and scales. G: Eggs of *H. binotata* are laid on bark near scales, where larvae hatch and seek out scales. Adapted from Simanton (1916).

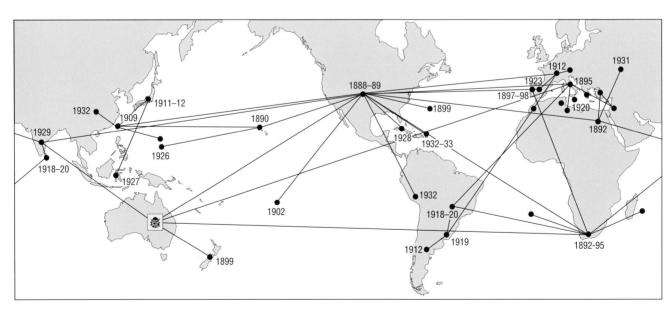

■ FIGURE 8-5. A map of the international movement of vedalia beetle during 1888 to 1933. This highly effective predator, originally from Australia, was subsequently introduced from California throughout the world to control cottony cushion scale infesting citrus.
Source: United States Department of Agriculture.

inches (1.5–3 mm) long. Its head is often more reddish or lighter colored than the dark wing covers and its underside is uniformly red or yellowish. The anterior underside and legs of the black lady beetle are black, with only the underside of the abdomen red. *Rhizobius forestieri* is about ¹⁄₁₀ to ¹⁄₆ inches (2.5–4 mm) long.

■

Twicestabbed Lady Beetle

The twicestabbed lady beetle (*Chilocorus stigma* =*C. orbus*) and other *Chilocorus* spp. are among the most common of the larger scale-feeding lady beetles. At least some *Chilocorus* spp. also feed on aphids, adelgids (aphidlike insects), and other soft-bodied insects. Many lady beetle larvae have spines, but *Chilocorus* spp. larvae are distinctive because they have prominent spines with multiple branches that look like spines on spines.

The twicestabbed lady beetle adult, which is shiny black with two red spots, resembles adults of at least three other species of lady beetles in California (*Axion plagiatum, Chilocorus kuwanae,* and one form of *Olla v-nigrum*). These species can be easily confused. *Chilocorus kuwanae* is the most recent of these species to arrive in California; it has been introduced throughout the United States to control euonymus scale, but also feeds on other armored scales, including San Jose scale (Bull et al. 1993).

MEALYBUG-FEEDING LADY BEETLES

The mealybug destroyer and many mostly smaller lady beetles (such as *Exochomus, Hyperaspis, Scymnus* spp.) are important predators of mealybugs. Some aphids, adelgids (aphidlike insects), scales, and other soft-bodied invertebrates are also common prey. Larvae of some species of these beetles are covered with whitish, waxy material, mimicking the appearance of their prey.

The mealybug destroyer (*Cryptolaemus montrouzieri*) is an important

A *Scymnus* spp. larva on a leaf infested with woolly aphids. Over a dozen *Scymnus* spp. occur in the United States, feeding on mites and soft-bodied insects such as aphids and mealybugs. Adults of most species are less than ⅙ inch (4 mm) long and range from entirely pale to entirely dark to dark with pale blotches. The pale, waxy *Scymnus* larvae resemble mealybug destroyer larvae, except *Scymnus* are much smaller than mature mealybug destroyer larvae.

predator of exposed mealybug species and certain other Homoptera such as the green shield scale. Both adult and larval lady beetles feed on all mealybug stages, but adults and young larvae prefer mealybug eggs and young nymphs. One lady beetle generation from egg to adult requires from 1 month at 70°F (21°C) to 2 months at 80°F (27°C). The beetle has about four generations a year and prefers 60 to 80% relative humidity.

The mealybug destroyer survives poorly over the winter and in most areas must be reintroduced in the spring to provide control. Some citrus growers purchase and release about 500 adult beetles per acre in spring to control mealybug species that feed openly. This predator is also released to control mealybugs in greenhouses and interiorscapes. Egg laying by females (fig. 8-6) is stimulated by presence of their prey's wax filaments (Merlin, Lemaitre, and Grégoire 1996). Mealybug destroyers are likely to reproduce and provide control only if released where mealybugs and their egg sacs are relatively abundant. A second generation of lady beetles (about 2 months after release) usually is necessary before progeny of the released beetles become abundant and provide control.

The adult mealybug destroyer is mostly dark brown or blackish with an orangish head and tail. Mealybug destroyer larvae are covered with waxy white curls and resemble mealybugs, except that the lady beetle larvae are more active, and mature larvae grow larger than mealybugs. The wax can be scraped off to reveal the pale, alligator-shaped beetle larva.

WHITEFLY-FEEDING LADY BEETLES

All stages of whiteflies are commonly preyed upon by *Delphastus* spp. lady beetles. Other lady beetles, including some *Chilocorus, Clitostethus, Hippodamia,* and *Scymnus* spp., feed on whiteflies at least occasionally.

Delphastus pusillus feeds on whiteflies throughout the southern two-thirds of the United States. Adults are shiny dark

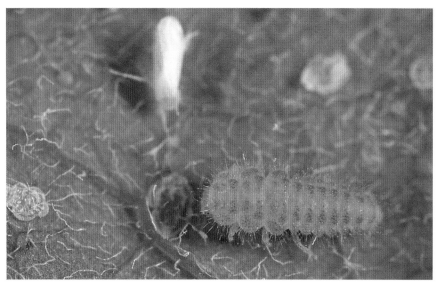

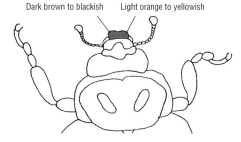

A Convergent lady beetle: Front topside

Females: Dark brown to blackish

Males: Light orange to yellowish

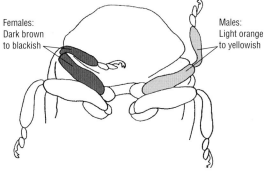

B Mealybug destroyer: Front underside

Females: Dark brown to blackish

Males: Light orange to yellowish

Like the adults, this tiny pale *Delphastus* larva feeds on all stages of whiteflies. This lady beetle, in combination with *Encarsia* spp. parasitic wasps, can provide better biological control than either *Delphastus* or wasps alone (Heinz and Nelson 1996, Heinz and Parrella 1994).

■ FIGURE 8-6. Color differences can generally be used to distinguish the sex of the convergent lady beetle (*H. convergens*) and the mealybug destroyer (*C. montrouzieri*): A: Convergent lady beetle females usually have a dark brown to blackish labrum mouthpart. The labrum of males is mostly light brown to orangish. The labrum can often be seen if beetles are viewed from above when they have their mouthparts extended fully forward to feed; however, in order to clearly see the labrum, it may be necessary to examine beetles from directly in front, such as by holding the beetle gently between your fingers or with a flexible tip forceps. B: Mealybug destroyers must be turned upside down to distinguish between males and females by examining their front (prothoracic) pair of legs. The front femora and tibiae of males are light orangish to yellow, whereas the front leg segments of females are dark brown to blackish. Mealybug destroyer progeny are important in providing biological control of mealybugs. When purchasing and releasing *Cryptolaemus*, determine the sex of a few beetles to make sure there are adequate numbers of females.

brown to black, sometimes with an orangish head or pale thoracic margins. Adult *Delphastus* have nine-segmented antennae with a terminal club formed from only the last segment.

MITE-FEEDING LADY BEETLES

Stethorus spp. lady beetles feed almost exclusively on tetranychid mites. These predators can be easily overlooked due to their tiny size, as illustrated by the adult on an almond pictured on page 2. Common North American species are *Stethorus picipes*, *S. punctum*, and *S. punctillum*.

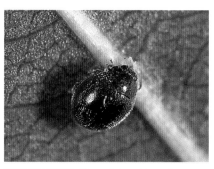

The adult spider mite destroyer (*Stethorus picipes*) is ⅟₁₆ inch (1.5 mm) long or smaller. It is shiny black with a very finely punctured surface covered with pale, minute hairs. Females lay tiny pale eggs among spider mite colonies. *Stethorus* may be most efficient in suppressing high spider mite populations (Tanigoshi and McMurtry 1977), but effective predation can occur too late in the season to prevent damage.

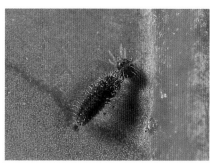

This dark gray to brownish spider mite destroyer larva is covered with numerous fine hairs. It is eating a citrus red mite, one of the half-dozen mites it can consume per day. Spider mite destroyer pupae are dark orangish to black and are covered with fine hairs.

PREDACEOUS GROUND BEETLES

Predaceous ground beetles (Carabidae) are medium to large soil-dwelling beetles. Most are predaceous, although some feed on seeds or organic litter (Lövei and Sunderland 1996). Over 2,500 species are known in North America. Their shape and color varies greatly. Adults are often black or dark reddish, although some species are brilliantly colored or iridescent. Most species have a prominent thorax that is narrower than their abdomen. Their long antennae have 11 segments and are not clubbed at the end. They have long legs, are fast runners, and rarely fly. Carabids resemble plant-feeding darkling beetles (Tenebrionidae). These groups can be distinguished as illustrated in figure 8-7.

Although most carabids dwell on the ground, some species climb bushes or trees to feed on caterpillars. Carabids can consume their body weight in food each day and are often abundant. They are believed to be important biological control agents, but their actual impact in most situations is undocumented. Little is known about most carabids in part because larvae feed out of sight in soil and litter, and adults usually hide during the day and stalk prey at night.

Both adults and larvae feed on soil-dwelling insect larvae and pupae (Riddick and Mills 1994) and other invertebrates such as snails and slugs (Asteraki 1993). Eggs are laid in moist soil, sometimes in specially molded mud cells attached to plants or stones. Larvae dwell in litter or soil and are elongate; the head is large and has distinct mandibles. Most species take about a year to complete the cycle from egg to adult (fig. 8-8) and adults live from 2 to 3 years. Because they are ground-dwelling, carabid populations may be enhanced by providing hedgerows or other undisturbed vege-

 LENGTH — Adult rufous carabids (*Calathus* spp.) are dark to bright reddish. They occur throughout the United States and feed primarily on insect larvae.

¾ INCH LENGTH — This adult predaceous ground beetle (*Calosoma* sp.) stalks its prey on soil, in litter, and sometimes in trees and shrubs.

⅞ INCH LENGTH — Plant-feeding darkling beetles (Tenebrionidae), as shown by this *Eleodes* sp., often resemble carabids. When disturbed, some tenebrionids raise their abdomen to a 45-degree angle and give off an offensive odor or distasteful liquid. Unlike predaceous ground beetles, darkling beetle antennae arise from beneath a distinct ridge on the head. These two groups can be distinguished by examining their hind legs at the underside of their abdomens (see fig. 8-7).

LENGTH — Carabid larvae have 10 well-defined, abdominal segments, a large head with prominent mandibles, and four-segmented antennae. The body tapers toward the rear with prominent appendages (cerci). An anal tube on the ninth segment is visible as a stubby pale protuberance between the cerci.

tation near crops. Tillage can reduce local density of carabids.

OTHER PREDATORY BEETLES

In addition to lady beetles and predatory ground beetles, over 3 dozen other families of Coleoptera contain predaceous species. Rove beetles (Staphylinidae), soldier beetles (Cantharidae), soft-winged flower beetles, and tiger beetles (Cicindelidae) are among the important predatory groups.

LENGTH — Adult soldier beetles are long and narrow, usually with an orangish head and thorax and dark wing covers. Adults such as this *Cantharis* sp. are often observed feeding on aphids or pollen on flowering shrubs and trees. Larvae are predaceous in the soil.

 LENGTH Adult rove beetles are often shiny black and have short folded wing covers that leave their abdominal segments visible from above. *Oligota oviformis* looks like this unidentified species; both adults and the yellow to orangish larvae are important mite predators, each eating about 10 to 20 mites per day. Pupae are dark orange to brownish. A similar-looking species (*Aleochara bilineata*) is an important natural enemy of onion maggot in the eastern United States (Hoffmann, Petzoldt, and Frodsham 1996).

 Most sap beetles (Nitidulidae) feed on decaying organic material, fungi, or plant sap. This predaceous sap beetle (*Cybocephalus californicus*) is an exception. These tiny, shiny black adults eat armored scales, such as the walnut scale shown here. Sap beetles can be distinguished from small lady beetles by the five segments on the end of each leg; lady beetles have three distinct segments (fig. 8-3).

LENGTH This Alaska tiger beetle (*Cicindela longilabris*) occurs throughout the western United States and Canada. It is iridescent black to brown or green with white marks on the wing covers. It is found primarily in sandy areas around conifer forests, where it preys on various insects.

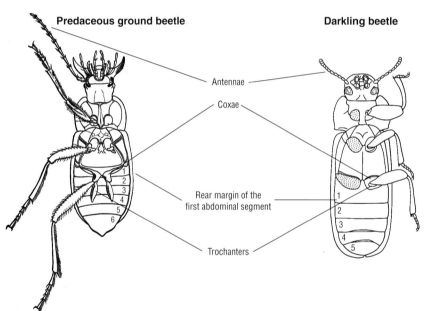

Predaceous ground beetle   **Darkling beetle**

Antennae

Coxae

Rear margin of the first abdominal segment

Trochanters

■ FIGURE 8-7. This diagram shows the underside of a predaceous ground beetle (Carabidae) and a darkling beetle (Tenebrionidae), with their inside legs removed and the abdominal segments numbered. Predaceous ground beetles can be distinguished from plant-feeding darkling beetles by examining where their hind legs meet their abdomen. Predaceous ground beetles, tiger beetles, and other suborder Adephaga beetles (most of which are aquatic) have enlarged basal segments (coxae and trochanters) on their hind legs; the hind leg coxae completely divide or cover at least the first abdominal segment. When you examine a darkling beetle's underside (and other suborder Polyphaga species) the rear margin of the first abdominal segment is entirely visible and the hind coxae and trochanters are not enlarged. In addition, unlike predaceous ground beetles, darkling beetle antennae arise from beneath a distinct ridge on the head. Predaceous ground beetle adapted from Packard 1876; darkling beetle adapted from Daly, Doyen, and Ehrlich 1978, reprinted with permission from H. V. Daly.

LENGTH The striped or twolined collops (*Collops vittatus*) is a soft-winged flower beetle. Adults have two broad bluish or black stripes on their reddish or orange wing covers. Males have enlarged basal antennal segments, as seen here. Twolined collops is common throughout the United States and Canada. Adults prey on aphids, spider mites, and eggs and larvae of alfalfa caterpillar and other moths in field and row crops, including corn and cotton (van den Bosch and Hagen 1966). The pinkish, soil-dwelling *Collops* larvae are at least partly predaceous, but may also scavenge dead organic matter.

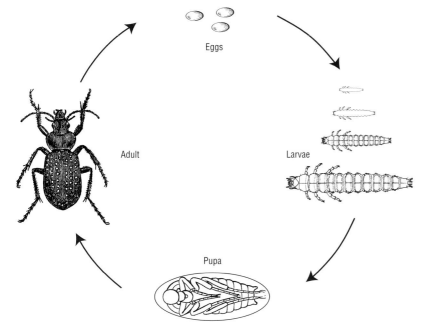

 FIGURE 8-8. Life cycle and stages of a typical predaceous ground beetle. All stages commonly occur only on the ground, with egg laying and pupation occurring under litter or just beneath the soil surface. However, adults of some species climb shrubs or trees to feed on caterpillars. *Sources*: Adult and pupa from Packard 1876; larvae from Peterson 1960, reprinted with permission of Helen H. Peterson.

TRUE BUGS

Species in the order Hemiptera (named Heteroptera by some specialists) are called true bugs or bugs. Although some people call any insect a bug, only Hemiptera are bugs. Although many bugs are plant feeders, many others, including most aquatic bugs, are predaceous.

Most predaceous Hemiptera do not specialize to the extent of many other predators; they are relatively general feeders, eating eggs, immature stages, and adults of a wide variety of insects and mites, including beneficial species. Both adults and nymphs are predaceous. At least 21 hemipteran families include predaceous species. Not discussed here are the 15 or more families of aquatic predaceous bugs (see Schuh and Slater 1995, Slater and Baranowski 1978, Usinger 1971), most notably the backswimmers (Notonectidae), shore bugs (Saldidae), water boatmen (Corixidae), and water striders (Gerridae). The most important terrestrial groups are described in table 8-2.

Hemiptera usually have thickened forewings with membranous tips. When true bugs are at rest, the dissimilar parts of their folded wings overlap, causing a characteristic X-shape on their back. Bugs have tubular mouthparts, which predaceous species can extend forward and use to impale their hosts and suck out the body contents. Mouthparts of most plant-feeding bugs can project only downward when the bug is viewed from above. Unlike plant-feeding species, some predatory bugs have front legs adapted for seizing prey (raptorial forelegs), which may be thickened and muscular with spines and sharp claws.

Eggs of some species (ambush bugs, assassin bugs, bigeyed bugs, and stink bugs) are laid openly in groups on plants. Other species (damsel bugs and minute pirate bugs) insert their oblong eggs into plant tissue so that eggs are virtually indistinguishable, often leaving only a round cap protruding. Eggs hatch into immatures called nymphs that gradually develop into adults without going through a pupal or cocoon stage (fig. 8-9). Nymphs resemble adults of the same species, except nymphs are

wingless, smaller, and often are a different color (Henry and Froeschner 1988, Slater and Baranowski 1978).

MINUTE PIRATE BUGS

Orius spp. are the most abundant minute pirate bugs in many habitats. They are common in field and row crops, including alfalfa, corn, cotton, small grains, soybeans, and tomatoes. Their prey includes aphids, mites, thrips, whiteflies, small caterpillars, and insect eggs. Commercially available *Orius* spp. are sometimes released in greenhouses, primarily against thrips (Coll and Ridgway 1995, van den Meiracker 1994). Unlike many bugs, which require 6 weeks or more to complete one generation, *Orius*

LENGTH ⊢ An adult minute pirate bug (*Orius tristicolor*) feeding on an aphid. A similar-looking species (*O. insidiosus*) is common in the eastern United States. *Orius* is one of the first and most common predators to appear outdoors in the spring.

LENGTH ⊢ Minute pirate bugs insert their eggs into plant tissue. Only the white caps of these *Anthocoris nemoralis* eggs protrude from this acacia leaflet; a smaller oblong, orange psyllid egg is entirely visible. This introduced pirate bug and an introduced lady beetle (*Diomus pumilio*) provide biological control of the acacia psyllid in California (Dreistadt and Hagen 1994). Like other minute pirate bugs, *A. nemoralis* is a general predator and also feeds on aphids, mites, thrips, and insect eggs.

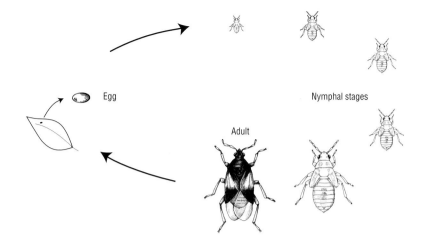

Egg

Nymphal stages

Adult

■ FIGURE 8-9. Life cycle and stages of a minute pirate bug. Only the tiny egg cap protrudes from the plant. All stages occur near their hosts on plants. Adults fly to seek out other prey and also occur on flowers and other nectar sources, where they feed on exuding sweet liquids. Adapted from drawings by Celeste Green in Smith and Hagen (1956).

TABLE 8-2. Common terrestrial predatory bug families.

Anthocoridae, minute pirate bugs or flower bugs

A

Most are predaceous on aphids, mites, thrips, psyllids, and insect eggs. Adults are about 1/12 to 1/5 inches (2–5 mm) long, oval, black or purplish with white markings, and have a triangular head. Nymphs are commonly pear-shaped and yellowish or reddish brown. *Orius* and *Anthocoris* are two common genera (Kelton 1978). Over 70 species in the United States.

Geocorinae subfamily of Lygaeidae, bigeyed bugs or seed bugs

B

Feed on both insects and seeds. Bigeyed bugs are oval, somewhat flattened, usually brownish or yellowish insects that stalk their prey. They have a wider head and more prominent bulging eyes than most other bugs. About 14 genera with over 200 species, including 20 *Geocoris* spp., north of Mexico.

Nabidae, damsel bugs

C

Predaceous on mites, aphids, caterpillars, leafhoppers, and other insects. Damsel bugs are mostly yellowish, gray, or dull brown, slender insects with an elongated head and long antennae. Adults look like small Reduviidae, up to 2/5 inches (10 mm) long. Nymphs sometimes look like ants. About 500 species worldwide.

Pentatomidae, stink bugs

D

Although most species are plant-feeding (mouth points downward), some are important predators (e.g., *Perillus*, *Podisus* spp.) and have hinged, tubular mouthparts that can point forward. They are oval or shield-shaped, commonly brownish, and usually are <2/5 inch (10 mm) long.

Phymatidae, ambush bugs

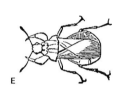

E

Ambush bugs wait motionless on plants and grab passing bees, flies, and wasps, prey often larger than themselves. Stout insects, 3/5 inch (15 mm) long or less, with broad, raptorial front legs and variable, cryptic colors, often black and brown, yellow, or green. About 300 known species.

Reduviidae, assassin bugs

F

All are predaceous. Some (e.g., kissing bugs, western bloodsucking conenose, western corsair) bite mammals; most eat only insects. Adults are blackish, reddish or brown, with a long narrow head, round beady eyes, and an extended, 3-segmented, needle-like beak. Adults usually are larger and have longer legs than most other bugs. Over 160 species in North America.

Emesinae subfamily of Reduviidae, thread-legged bugs

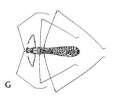

G

All are predaceous on various small insects. Adults have an elongate, slender body with long, threadlike legs and are often brownish, resembling walking sticks. Many species are tropical and are not well known; at least several hundred species may exist.

Sources: Slater and Baranowski 1978, Schuh and Slater 1995 (A-C *from* Smith and Hagen 1956, by Celeste Green; F, G *from* Zimmerman 1948, reprinted with permission from University of Hawaii Press; D, E *from* Packard 1876).

LENGTH

Yellowish *Orius* nymphs like the one shown here emerge from oval eggs laid in plant tissue. Older nymphs may be darker but also have red eyes.

develop from egg to adult in less than 3 weeks at 77°F (25°C).

◼

LOOK ALIKE BUGS

Certain groups of plant-feeding pest bugs resemble some predatory bugs. For example, plant-feeding chinch bugs (Lygaeidae) may be confused with predatory bigeyed bugs, which are in the same family. Like most other plant-feeding Lygaeidae, chinch bugs are more slender and have smaller eyes than bigeyed bugs. Minute pirate bugs and bigeyed bugs can be mistaken for plant bugs (Miridae). Most plant bugs are plant feeders and many are pests. However, at least three genera of mirids (*Deraeocoris, Hyaloides,* and *Phytocoris*) include predaceous species that feed on aphids, caterpillars, insect eggs, mites, psyllids, and other small arthropods (Alomar and Wiedenmann 1996, Kelton 1982).

LENGTH

Plant bugs (Miridae) are larger than most similar-looking minute pirate bugs or bigeyed bugs. This important pest species, *Lygus hesperus,* has a distinct yellow or greenish triangle on its back. Unlike other bugs, the forewings of plant bugs have only one or two closed cells in the membranous portion (tip), and they have a distinct flap (cuneus) near the middle of the forewing outer margin. Lygus bugs also have longer antennae than minute pirate bugs or bigeyed bugs.

◼

BIGEYED BUGS

LENGTH

Bigeyed bugs such as this adult *Geocoris punctipes* are common in low-growing plants, many field and row crops, and on the ground. Other California species are *G. atricolor,* which is mostly shiny black, and *G. pallens,* which is yellowish with reddish brown spots.

LENGTH

Bigeyed bug nymphs such as these *G. punctipes* are oval and somewhat flattened. Their eyes are widely separated, giving them a wide field of vision for spotting their prey, which includes other bugs, flea beetles, spider mites, insect eggs, and small caterpillars.

◼

ASSASSIN BUGS

¾ INCH
LENGTH

The leafhopper assassin bug (*Zelus renardii*) is common throughout California. Like most assassin bugs, the adult (left) is a poor flyer. The adult and nymph (right) stalk or lay in wait for their prey, which they inject with venom. Despite being called the leafhopper assassin bug, this predator feeds not only on leafhoppers but on a variety of small to medium-sized insects.

LENGTH

Plant-feeding chinch bugs (Lygaeidae) look like this false chinch bug (*Nysius raphanus*) and may be confused with bigeyed bugs. Chinch bugs are more slender and have smaller eyes than bigeyed bugs.

LENGTH

Bigeyed bug eggs are laid singly, usually on the leaf surface. Eggs are oblong and pale colored and develop a reddish eyespot shortly after being laid.

LENGTH

Zelus spp. assassin bug eggs are barrel-shaped, dark brown with a white cap, and are laid openly in groups.

The forelegs of this spined assassin bug (*Sinea diadema*) are thickened and have sharp barbs (raptorial forelegs) for holding prey. This polyphagous predator is widely distributed throughout the United States.

The spined assassin bug (*Sinea diadema*) lays its eggs in a cluster. Each egg has an umbrella-like cap.

Thread-legged bugs are more slender and have a more elongated thorax than other predaceous Reduviidae. This *Emesaya brevipennis* and related species resemble walking sticks.

OTHER PREDATORY BUGS

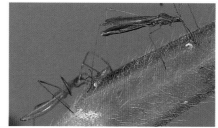

Unlike the similar-looking thread-legged bugs, this spined stilt bug (*Jalysus wickhami*, Berytidae) does not have a greatly elongated thorax and its front legs are not raptorial. The spined stilt bug is omnivorous, feeding on both insects and plant sap. Adults and nymphs occasionally damage tomatoes by puncturing and killing fruit stems and blossoms. This species also eats caterpillar eggs and soft-bodied prey such as aphids (Jackson and Kester 1996, Wheeler and Henry 1981).

Damsel bugs, such as this *Nabis* sp. adult and nymph, occur in row crops and low-growing plants, such as alfalfa, and in shrubs and some orchard trees. Damsel bugs often appear later in the season than other species and move rapidly when disturbed. Their prey includes aphids, other bugs, leafhoppers, spider mites, thrips, and small caterpillars.

This twospotted stink bug (*Perillus bioculatus*) has attacked a Colorado potato beetle larva. Although many stink bug species are pests, important predators are found in several genera (such as *Perillus* and *Podisus*). Predatory stink bugs are sometimes called soldier bugs.

PREDATORY FLIES

Flies, order Diptera, are mostly small, soft-bodied insects distinguished from other groups by their single pair (rather than two pairs) of wings. At least 20 fly families have species that are predaceous as larvae or adults; some important groups are summarized in table 8-3.

The most frequently observed predaceous flies are flower flies or hover flies (Syrphidae). Aphid flies (Chamaemyiidae) and predaceous midges (Cecidomyiidae) are other predatory flies with habitats similar to that of flower flies. Adults in these three groups are not predaceous; they eat pollen and nectar. Adults lay their oblong eggs singly or in scattered groups on plants near colonies of mites or soft-bodied insects such as aphids, which are the most common prey of the larval stages. The predaceous maggotlike larvae (fig. 8-10) molt through three stages before pupating into adults.

SYRPHID FLIES

Adult flower flies (Syrphidae), also called syrphid flies or hover flies, are large to medium-sized flies. Syrphids are the most common group of predaceous flies. Hundreds of species occur around farms and gardens in North America; about 6,000 species are known in the world.

Most adult syrphids eat nectar and pollen and can be important plant pollinators. Adults commonly hover around flowers and many species resemble honey bees. Larvae of many syrphids feed almost exclusively on insects, mostly Homoptera, especially aphids. Larvae of a few species are pests, such as *Merodon equestris* and *Eumerus* spp., which bore into bulbs of narcissus and related plants. Some syrphids are innocuous, including those that live in nests of ants, bees, and wasps. Certain syrphids feed on fungi or decaying organic matter. For example, *Eristalis* spp. have rat-tailed maggots that live in manure and sewage

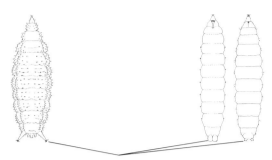

Projecting anal spiracles

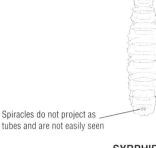

Spiracles do not project as tubes and are not easily seen

APHID FLY

Larvae have two anal spiracles (small tubes) protruding from their rear. These tubes are widely spaced and relatively long.

PREDACEOUS MIDGE

Larvae have two anal spiracles that are shorter and closer together than on aphid midges.

SYRPHID FLY

Larvae lack distinctly projecting anal spiracles. They usually have a more opaque cuticle or skin through which internal organs can be seen. Syrphid larvae are usually larger and more variously colored or patterned than aphid flies or predaceous midges.

■ FIGURE 8-10. Maggotlike larvae of three fly families that eat aphids, mealybugs, scales, and other soft-bodied insects. These aphid fly, predaceous midge, and syrphid fly larvae can be dis-tinguished by the two small breathing tubes (anal spiracles) at their rear. Adapted from Peterson 1960, reprinted with permission of Helen H. Peterson.

LENGTH

Adult syrphids commonly have black and yellowish abdominal bands. Although many syrphid species resemble honey bees, they cannot sting. Syrphids can be distinguished from bees by their one pair of wings and distinctive manner of hovering in flight.

LENGTH

The large hover fly larva (*Scaeva pyrastri*) is light green with a white line down its back. At first glance this beneficial might be mistaken for a pest caterpillar. Larvae of some syrphids and all aphid flies and predaceous midges are much smaller than most mature caterpillars and can easily be overlooked among a colony of aphids.

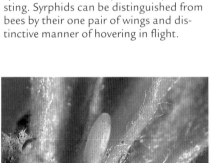

LENGTH

Syrphid eggs are laid singly on their sides near aphids or other prey. Eggs are oblong, usually whitish to gray, and have a surface covered with crossing strands or lines. Brown lacewing eggs look similar, except they have a smoother surface and a tiny knob projects from one end of the egg.

LENGTH

In a futile effort to repel the *Metasyrphus* sp. flower fly larva eating it, this rose aphid is secreting from a cornicle a droplet of noxious fluid, which also serves as an alarm pheromone warning other aphids of danger. Upon maturity, syrphid larvae form teardrop-shaped to oblong pupae that are usually green to dark brownish. Pupae are found attached to plants near colonies of aphids or where aphids occurred.

ponds; larvae mature into adults called drone flies, which make a loud buzzing sound that imitates bees. Larvae of some syrphids (*Toxomerus* spp.) that eat mostly pollen may also consume aphids or spider mites.

Common aphidophagous syrphid species in California are found in at least 14 genera, including *Allograpta, Eupeodes, Metasyrphus, Scaeva,* and *Syrphus.* Adults are active throughout the year in mild-climate coastal areas and can have several generations each year (fig. 8-11). In interior areas of the state, syrphids are abundant during spring and early summer, but most species become uncommon during very hot or cold weather. Adult syrphids are strong fliers and good searchers.

■

PREDACEOUS MIDGES AND APHID FLIES

Predaceous midges (Cecidomyiidae) and aphid flies (Chamaemyiidae) are smaller than most syrphids but have a similar life cycle (fig. 8-12). Aphid fly larvae are beneficial predators of aphids, mealybugs, scales, and other soft-bodied insects such as pear psylla. Predaceous midges are in the same family as many gall-making pest species (Gagné 1989), but predaceous midges feed on mites and aphids and other small, soft-bodied insects.

TABLE 8-3. Common predatory fly families.

Asilidae, robber flies

All are predaceous on insects as both adults and larvae. Adults often catch large prey in flight. Larvae live in soil and litter, eating primarily other insect larvae. Adults have a stout thorax and strong legs, a bearded face, and a hollowed head on top between their eyes. About 800 species in North America.

A

Cecidomyiidae, gall midges or predaceous midges

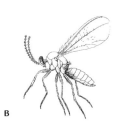

Many gall midge larvae feed on plants and cause tissue to distort into galls; larvae of predaceous species eat mites and soft-bodied insects such as aphids. Adults are tiny, slender, delicate flies with long slender legs and antennae. The predaceous larvae may be confused with aphid flies, but the two protruding anal spiracles on midge larvae are shorter and closer together. *Aphidoletes* spp. are released to control aphids in greenhouses.

B

Chamaemyiidae, aphid flies

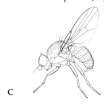

Larvae feed on aphids, mealybugs, and soft scales. Larvae are maggotlike with two prominent, widely spaced tubular anal spiracles protruding from their rear ends. Adults are small, chunky, densely hairy, grayish flies.

C

Chaoboridae, phantom midges

Larvae are predators of mosquito larvae, small crustaceans, and other small aquatic species. The nearly transparent ("phantom") aquatic larvae look like mosquito larvae. Unlike mosquitoes, phantom midge larvae have stout spines around their mouth. Adults are common near water and look like mosquitoes, but their tubelike mouth is short while mosquitoes have a long proboscis. About two dozen species in the United States.

D

Empididae, dance flies

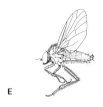

All are predators as both adults and larvae. The pale, cylindrical larvae feed on bark beetles under bark or on various prey in soil, litter, or water. Adults stalk small insects on bark or flowers. Adults have a large thorax and tapering abdomen. They often swarm and move up and down, as if dancing. Most are minute. Over 700 species in North America.

E

Mydidae, mydas flies

Larvae prey on beetle larvae and other insects, often in sand or decaying wood. Adults reportedly are predaceous, but their habits are not well known. Adults are large, elongate, and often brightly colored. Larvae are pale with a dark head, somewhat flattened, and up to 1⅕ inches (55 mm) long. Of the over 60 species that occur in North America, most are rarely seen.

F

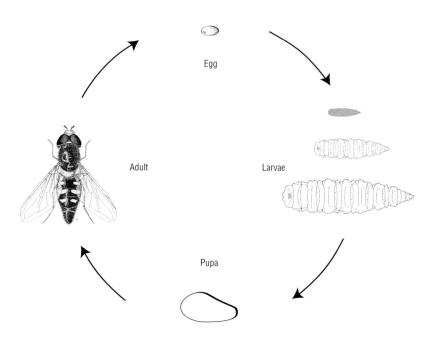

■ FIGURE 8-11. Syrphid larvae hatch from oblong, whitish to gray eggs laid singly on their sides near aphids. The predaceous larvae molt through three instars before forming a greenish to dark brownish oblong to pear-shaped pupa, usually on the plant, from which the adult emerges. Adults resemble bees and can be identified using the keys in Vockeroth (1992) or the other references included there (larvae *from* Peterson 1960, reprinted with permission of Helen H. Peterson; adult *from* Cole 1969, by C. S. Papp, reprinted with permission from University of California Press).

TABLE 8-3. Common predatory fly families *(continued)*.

Rhagionidae, snipe flies

Adults are predaceous on insects and are common in woods and moist areas. The whitish cylindrical larvae are mostly scavengers in moist soil, litter, or water, but may sometimes prey on immature insects. Adults have a relatively long, tapered body. Adults and larvae are ½ to 1 inches (12–24 mm) long. Over 100 known species.

G

Sciomyzidae, marsh flies

Larvae are aquatic and eat snails and slugs. Adults swarm around moist areas and are small to medium, usually yellowish or brown, with antennae projecting forward. The cylindrical, pale to brownish larvae are ⅓ to ½ inch (8–12 mm) long and have distinct spiny abdominal rings. About 150 species in North America.

H

Sources: Borror, De Long, and Triplehorn 1981; Daly, Doyen, and Ehrlich 1978; Peterson 1960 (A, C–J *from* Cole 1969, G, I, by C. S. Papp, reprinted with permission from University of California Press; B *from* Quayle 1932).

Syrphidae, syrphid flies, flower flies, or hover flies

Most adults eat pollen and nectar. Some larvae eat fungi, but many are important predators. Adults often have black and yellowish bands on their abdomen, commonly hover around flowers, and look like honey bees, but do not sting. Aphids and other soft-bodied insects are primary prey of larvae. Unlike aphid fly or predaceous midge larvae, flower fly larvae usually have a more opaque skin through which internal organs can be seen, are often more tapered, usually larger, and are variously colored or patterned. About 1,000 species in North America.

I

Therevidae, stiletto flies

Larvae are predaceous on other insect larvae in soil and decaying wood. Adult feeding habits are largely unknown. Adults occur in dry, open areas such as meadows; they are medium sized, usually hairy, and often have a pointed abdomen. The whitish larvae are very slender and about ⅘ to 1⅕ inches (20–30 mm) long. About 130 North American species are known.

J

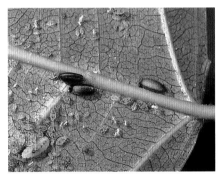

LENGTH

The mostly pale aphid fly larva (bottom left) and orangish pupae (along the leaf vein) have two widely spaced anal spiracles projecting prominently from their rear ends. These chamaemyiids are predators of aphids, mealybugs, scales, and other soft-bodied insects. For example, *Leucopis* spp. prey on pear psylla in pear orchards.

LENGTH

Predaceous midge adults, such as this *Aphidoletes aphidimyza*, are delicate flies with long slender legs. They often stand with their antennae curled back over their head. Adults are attracted to aphid honeydew, feed on honeydew and nectar, and are not predaceous.

LENGTH

Predaceous midge larvae, such as this *Aphidoletes aphidimyza*, have two projecting anal spiracles (small tubes) relatively close together (sometimes touching) at their rear ends. *Aphidoletes* is commercially available and is released in greenhouses for aphid control (Malais and Ravensberg 1992). This predator apparently thrives under high humidity.

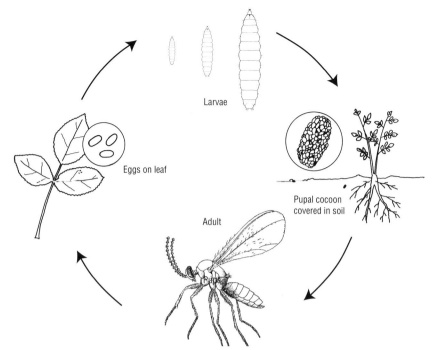

Yellow-legged paper wasps (*Mischocyttarus flavitarsis*) on their nest under a building eave.

LENGTH

A *Polistes* sp. yellowjacket with a caterpillar it captured in cotton.

LENGTH

■ FIGURE 8-12. The life cycle of a predaceous midge (*Aphidoletes aphidimyza*). Eggs are orangish red, oval, and only about ⅟₈₀ inch (0.3 mm) long. Larvae develop through three instars, are pale yellow to red or brown, and at maturity are about ⅟₁₀ (2.5 mm) long. Pupae are orange to brown, about ⅟₁₂ (2 mm) long, and occur beneath plants in litter where they may form cocoons made from soil particles, excrement, and aphid cast skins. Adults are delicate flies with long slender legs. *Sources:* Malais and Ravensberg 1992 (larvae *from* Peterson 1960, reprinted with permission from Helen H. Peterson; adult *from* Quayle 1932).

LENGTH

This *Feltiella* sp. predaceous midge larva is eating a spider mite egg. It also preys on the mite nymphs and adults. Larvae pupate in silken cocoons among colonies of their prey. This predator is important in strawberry fields.

PREDATORY WASPS AND ANTS

■

Ants and yellowjackets are the best known-predatory wasps (order Hymenoptera). Table 8-4 lists common wasp families that are predaceous during the adult stage or are parasitic or predaceous as larvae. Females live alone or in groups with other wasps. Wasps usually nest in the ground, in hollowed out plant parts, or in paperlike structures attached to buildings or other objects.

Adults have chewing mouthparts and feed on many different individual prey throughout their lifetime. Adult wasps also capture prey to feed their larvae. Many species paralyze their hosts with venom before laying an egg on it or car-

rying it back to their nest. During its development into an adult, a wasp larva may be parasitic, consuming a single prey or predaceous, consuming many prey. Ants, which are wingless wasps, have a greater variety of feeding habitats than other wasps. Invertebrates, plants or plant by-products (such as honeydew or seeds), and fungi provided by worker ants as regurgitated liquid are eaten by ant larvae.

Certain predatory *Polistes* spp. wasps have been used in biological control projects. Unlike some accidentally introduced yellowjacket species, which often annoy people and attack native invertebrates and vertebrates, many native predatory wasp species tend not to attack people and are generally considered to be beneficial.

TABLE **8-4.** Common predatory wasp families.

Formicidae, ants

A

Depending on species, ants are predaceous, feed on plants or fungi, or consume all of these. Some ants disrupt biological control by attacking predators and parasites. Ants are social insects that live in colonies containing reproductive queens, sterile female workers, and short-lived, winged males. Some species of ants sting or discharge irritating chemicals. Ants are sometimes confused with termites; ants can be distinguished by their distinctly elbowed antennae and the narrow constriction between their abdomen and thorax. Ants are the most abundant of all insects in terms of biomass and number of individuals; in tropical areas there are often several million ants per acre. About 600 species in Canada and the United States.

Mutillidae, velvet ants

B

Larvae of most species are external parasites of wasps and bees, especially ground-nesting bees; a few species attack beetles and flies. Females seek out and oviposit in hosts' nests. Adults are densely covered with fine, sometimes colorful hairs, but otherwise they resemble ants in shape and because mutilids walk rapidly on the ground. Females are wingless, and males are winged. Adults may make a distinct squeaking noise if held; females have a painful sting. About 500 species in North America, most in open or arid areas of the South and West.

Pompilidae, spider wasps

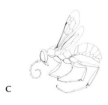

C

Larvae are parasites of spiders and tarantulas. Females capture and paralyze the host, construct a nest around it, and lay one egg on each spider or tarantula. All are solitary wasps. Some species paralyze spiders and carry them back to a nest already constructed elsewhere. A few species attack the spider or tarantula in its own burrow and lay their egg there. Adults often seek prey by running erratically along the ground while twitching their wings. Adults have a steel blue or black body with smoky or yellowish wings. They are ⅓ to 2 inches (8–50 mm) long with long legs and antennae. About 300 species in North America.

Sphecidae, digger wasps, mud daubers, and sand wasps

D

Some sphecids apparently feed their larvae with virtually any type of available invertebrate, but many species prey on only one type of insect or spider. Adults of virtually all orders of insects are attacked by one or more specialized sphecid species. Most nest in the ground. All species are solitary, although many individuals may nest in one area. Adults are about ½ to 1½ inches (12–37 mm) long. Sphecids can be distinguished from other wasps by their pronotum; the rear margin is straight when viewed from above. Most sphecids also have a raised constriction between their pronotum and mesoscutum (the top of the second or middle segment of the thorax, fig. 7-2), which looks like a distinct collar. Over 1,100 species in North America.

Vespidae, hornets, paper wasps, and yellowjackets

E

Although adults often eat caterpillars, they feed their larvae a variety of insects, including beetles, flies, true bugs, and other wasps. The most notable vespids are about three dozen common social species that display little visible difference between workers and the somewhat larger queen. Adults are black with conspicuous yellow or white markings. When viewed from above, the rear margin of the pronotum is strongly U-shaped. When at rest, these wasps typically fold their wings lengthwise. Instead of stinging, vespids usually kill prey by biting it with their mandibles. Their stinger is used to deter vertebrates; many other insects mimic these wasps to fool birds and other predators into avoiding them. Colonies exist only for one season in cold-winter areas, starting anew from an overwintered queen. In mild-winter areas, a colony can grow for years and become very large. Several hundred mostly solitary species in North America.

Note: These families include species that are predaceous during the adult stage, are parasitic or predaceous as larvae, or both.

Sources: Daly, Doyen, and Ehrlich 1978; Powell and Hogue 1979 (A–E *from* Goulet and Huber 1993, reproduced with the permission of the Minister of Public Works and Government Services Canada, 1997).

ANTS

Ants (Formicidae) are the most abundant predators in many natural areas, greatly influencing populations of invertebrates and other organisms. These predatory wasps differ from those in many other hymenopteran families in that ants are usually wingless. An exception is those born to mate and disperse; they lose their wings after leaving their old colony and establishing a new colony.

Native gray ants (*Formica* spp.) are beneficial predators in forests (Float and Whitham 1994, Weseloh 1994, 1995). Fire ants (*Solenopsis* spp.) are major predators of cotton pests in the southern United States (Sterling et al. 1984). Nests of weaver ants (*Oecophylla* spp.) are gathered, sold, and placed in selected citrus orchards in southern China because these ants prey on harmful insects, preventing damage to fruit (Huang and Yang 1987); this earliest-known example of biological control has been practiced for at least 1,700 years. *Oecophylla longinoda* and *O. smaragdina* reportedly are beneficial predators in certain tropical crops throughout Southeast Asia

and the Pacific Ocean islands (Way and Khoo 1992).

There has been relatively little research on using ants for biological control in the United States. Although native species such as California gray ants or field ants (*Formica* spp.) may be important predators of peach twig borer in peach orchards and other pests in certain situations, ants are generally not considered to be beneficial in California agriculture. Especially problematic are the Argentine ant and other introduced species that disrupt biological control. Some ants directly damage young tree bark and certain crops. For example, the introduced pavement ant and the native southern ant or California fire ant chew almonds. See chapter 6 for information on ant control and disruption of biological control by ants. To identify ant species, consult *Ants of California with Color Pictures* (Haney, Phillips, and Wagner 1987), *The Ants* (Hölldobler and Wilson 1990), and *Illustrated Key to Ants Associated with Western Spruce Budworm* (Shattuck 1985).

LACEWINGS, DUSTYWINGS, AND SNAKEFLIES

Lacewings, dustywings, and snakeflies belong to the order Neuroptera. Insects in this order are mostly predaceous, including dobsonflies (Corydalidae), alderflies (Sialidae), antlions (Myrmeleontidae), and related families not discussed here. Adult Neuroptera ("nerve-winged") are soft-bodied and have four membranous wings with many-branched veins. Adults have chewing mouthparts and most are predaceous, although adults of some species feed only on pollen and nectar or do not eat at all. Larvae of most species are flattened predators that are ⅛ to ⅘ inch (3–20 mm) long, with prominent jaws (Tauber 1991).

GREEN LACEWINGS

Green lacewings (Chrysopidae) are very common predators in agricultural, garden, and landscape habitats (fig. 8-13). Lacewing larvae are sometimes called aphidlions because they often feed on aphids. However, lacewings also prey on mites and a wide variety of small insects, including caterpillars, leafhoppers, mealybugs, psyllids, whiteflies, and insect eggs. All lacewing larvae are predaceous, but adults of many species of green lacewings are not predaceous. For example, adult *Chrysopa* spp. feed on insects, honeydew, nectar, and pollen; *Chrysoperla* spp. adults feed only on honeydew, nectar, and pollen (Canard, Séméria, and New 1984). Green lacewings are commercially available and are among the most commonly released predators.

Releasing Green Lacewings. *Chrysoperla* spp. can be purchased for release as eggs, adults, or larvae; eggs are the most common and least expensive stage that is commercially available. The lacewings are usually shaken from the commercial shipping container onto infested plants; modified pesticide sprayers are used for large-scale releases. Although widely used in some greenhouses and row crops, the effectiveness of releasing lacewings has been variable and there is little information on using them in most situations. Conserving resident beneficials is often likely to be more effective and economical.

Most early control programs using lacewings released *Chrysoperla carnea* eggs (Ridgway and Murphy 1984, Tulisalo 1984). Release of *Chrysoperla carnea* was ineffective against cotton aphid (Rosenheim and Wilhoit 1993), and release of three *Chrysoperla* spp. provided variable results against grape leafhopper (Daane et al. 1993, 1996). *Chrysoperla rufilabris* release has been found to control conifer aphids in a seedling nursery (Ehler and Kinsey

Ants are generally considered to be pests in California agriculture. However, this California gray ant or field ant (*Formica aerata*) can be a beneficial predator in certain situations. This ant has captured a peach twig borer larva. This species will also tend and protect honeydew-producing pests such as aphids and soft scales.

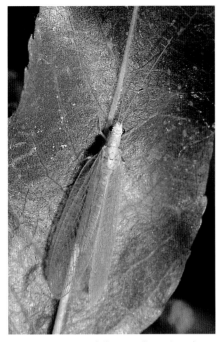

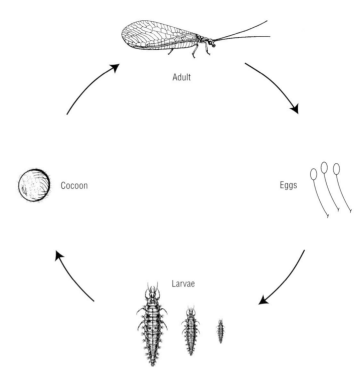

■ FIGURE 8-13. Green lacewing life cycle and stages. Tiny oblong eggs, each with a silken stalk, are attached to plants. Larvae develop through three instars. Loosely woven, spherical, silken cocoons are attached to plants or in crevices, such as under loose bark. Adults have golden eyes, slender bodies, and are named for their green, lacy, wing veins. Average development time from egg to adult is about one month. *Sources:* Tassan and Hagen 1970 (adult and larvae *from* Smith and Hagen 1956, by Celeste Green).

¾ INCH

LENGTH

Adult green lacewings have golden eyes and slender green bodies. They are named for their prominent wings, which have green, netlike or lacy veins. These night-flying insects are often seen when drawn to lights. During the day, adults can be observed flying if their resting place on foliage is disturbed, such as by branch-beat sampling. The species of adult lacewings can be distinguished only by an expert examination of their genitalia.

LENGTH

Lacewing larvae pupate in a loosely woven, spherical, silken cocoon attached to plants or under loose bark. Although lacewings can overwinter as pupae, all stages except eggs can be present throughout the year in areas with a mild climate. Prolonged cool weather and short day length causes adults and larvae to be largely inactive for weeks.

LENGTH

Lacewing larvae, such as this third-instar *Chrysoperla rufilabris* eating a rose aphid, have long, curved, tubular mandibles for puncturing and sucking the fluids out of their prey. Larvae are flattened, tapered at the tail, have distinct legs, and look like tiny alligators. Larvae are usually pale with darker markings (fig. 8-14) and develop through three instars. Larval markings can be used to distinguish species (Tauber 1974).

LENGTH

Tiny, oblong, green lacewing eggs, each with a silken stalk, are attached to this aphid-infested asparagus. Depending on the species, eggs are laid singly or in clusters. Eggs are green when laid, then darken before hatching. Eggs hatch about 4 days after being laid, and larval development through 3 instars takes about 10 days when temperatures average about 77°F (25°C).

LACEWING SPECIES	FIRST INSTAR	SECOND INSTAR	THIRD INSTAR
Chrysoperla carnea Occurs in vegetation throughout the United States, but may be more common in low-growing habitats. Adults are not predaceous. Eggs are laid singly. Is commercially available.			
Chrysoperla comanche Arboreal, often occurs in managed habitats. Adults are not predaceous. Eggs are laid singly.			
Chrysopa coloradensis Apparently prefers shrub and tree habitats. Adults are predaceous.			
Chrysoperla rufilabris Occurs in most types of vegetation, but may be more common in tree and shrub habitats. Adapted to more humid environments; common in the midwestern and eastern United States, but not naturally common in the West. Adults are not predaceous. Eggs are laid singly. Is commercially available.			
Chrysopa nigricornis Common throughout the United States; apparently prefers arboreal habitats. Adults are predaceous and require relatively high host density before laying eggs. Eggs are laid in groups.			
Chrysopa oculata Appears to be relatively uncommon. Adults are predaceous.			

■ FIGURE 8-14. Head markings on the three instars of several common green lacewing species. When purchasing and releasing green lacewings, identifying the species present can help you determine whether the lacewings present are the species you released or larvae of naturally occurring species.
Sources: Adapted from Tauber 1974, reprinted with permission of the Entomological Society of Canada; adapted from Zheng n.d. unpublished.

1995), silverleaf whiteflies on green-house hibiscus (Breene et al. 1992), and Colorado potato beetle larvae in caged plants (Nordlund, Vacek, and Ferro 1991), but releases were ineffective against the green apple aphid (Grass-witz and Burts 1995).

How lacewings are released can greatly affect the program's success. Mechanical methods are being developed to improve release efficacy and economics (Gardner and Giles 1996a, 1996b). If releases are planned, protect eggs from predation and avoid injuring the eggs or larvae. Releasing adequate numbers of lacewing larvae may be more likely to control pests than dispersing eggs. Eggs can be held at room temperature until larvae begin hatching, then distributed. Augmentation is more likely to be effective if several weekly releases are made beginning when prey are first observed but before the pests become abundant. High pest populations prior to release can be reduced by applying a nonpersistent insecticide, such as insecticidal soap or oil, followed by predator releases the next day. Because adult lacewings tend to disperse before laying eggs (Duelli 1980), only the insects released as immatures (and probably not their progeny) are likely to provide any control. See the discussion of releasing natural enemies effectively in chapter 6 before introducing lacewings.

BROWN LACEWINGS

Brown lacewings (Hemerobiidae) are predaceous as both larvae and adults. They feed on mites and soft-bodied insects, especially aphids, mealybugs, scales, and whiteflies. Brown lacewing biology is similar to that of green lacewings, except that brown lacewings seem to prefer cooler temperatures and appear to be less common. Adults of at least 13 different *Hemerobius* spp. brown lacewings occur in North America and can be identified using the keys in Kevan and Klimaszewski (1987). The distribution and biology of many of these is poorly known.

LENGTH

Brown lacewing eggs, such as this *Hemerobius* sp., are oblong and are laid singly on their side on plants without any attaching stalk. Although brown lacewing eggs look similar to syrphid eggs, brown lacewing eggs have a tiny knob projecting at one end and have a smoother surface than syrphid eggs, which appear to have fibers or lines crisscrossing their surface.

LENGTH

Adult brown lacewings, such as this *Hemerobius* sp., look like green lacewings, except that brown lacewings are typically about one-half as large and are light brown. Adults emerge from flat, white, silken cocoons.

OTHER NEUROPTERA

Dustywings (Coniopterygidae) and snakeflies (suborder Raphidioptera, family Raphidiidae) are also in the lacewing order Neuroptera. Dustywings are often said to be rare, but they may simply be overlooked because they are small. Over 20 species are known in California alone. Snakeflies are more commonly observed than dustywings because snakeflies are about 10 times larger than dustywings. Although snakefly adults and larvae are predaceous, their importance in biological control is unknown.

LENGTH

Brown lacewing larvae such as this *H. pacificus* are more slender and have a smaller head and jaws than green lacewings. Unlike green lacewings, brown lacewing larvae often move their head rapidly from side to side when seeking prey. Mature brown lacewing larvae (second and third instars) lack the trumpet-shaped lobe (empodium) that green lacewing larvae have between the claws on the end of their legs (fig. 8-15).

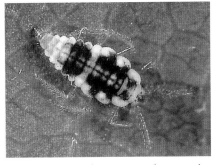

Dustywing larvae have a strongly tapered abdomen. When feeding on mites, they eat about 250 mites while developing through 3 larval stages. Dustywings occur mostly in shrubs and trees, where they pupate in an inconspicuous, flat, white, silken cocoon, often on the undersurface of leaves.

LENGTH

Adult dustywings are tiny insects with long antennae and prominent eyes. They are named for the whitish powder covering their wings. Adults lay minute, oblong eggs on foliage among colonies of mites or small insects. Dustywing larvae feed on all mite stages and virtually any tiny insect they can capture.

Behavior-modifying volatile chemicals (semiochemicals) are important in biological control. Many natural enemies locate their prey by detecting semiochemicals produced by their prey, the host plant, or a combination of both. Adult green lacewings require at least two different volatile chemicals before they are attracted to land in a crop. One necessary lacewing attractant is plant-produced; the specific chemical (a synomone) varies according to the plant species. In cotton, the plant-produced green lacewing attractant is caryophyllene, produced only while young bolls (squares) are developing.

The other necessary lacewing attractant is tryptophan, a product of the breakdown of honeydew produced by insects that suck plant-juices (Hagen 1986). This semiochemical, produced by one organism and benefiting another species, is called a *kairomone*. Honeydew is the principal food source and a host-location attractant for many beneficial species, such as aphid parasites (Grasswitz and Paine 1993). Natural enemies are also attracted to volatile chemicals emitted by insect-damaged plants (Vaughn, Antolin, and Bjostad 1996), constituents of the covers of scale insects (Hare and Luck 1994), insect excrement (frass), alarm pheromones secreted by aphids through their cornicles (Grasswitz and Paine 1992), and pheromones or sex attractants of prey of natural enemies.

Some beneficials can be attracted to crops (even when pests are absent) by applying the proper kairomone or synomone or both. For example, wheast (a by-product of cheese making) is a milk protein and yeast mixture that contains tryptophan and attracts green lacewings. Because lacewings also require sugar in their diet and to stimulate egg laying, sucrose (normally contained in insect honeydew) can be added to the wheast to create a food spray or artificial honeydew for application to crops (Hagen, Sawall, and Tassan 1971). Application of sucrose alone to plants sometimes increases the number of natural enemies present, including lady beetles, predaceous bugs, syrphid flies, and parasitoids (Evans and Swallow 1993). Although artificial foods containing sugars, yeasts, and proteins are commercially available to attract or improve reproduction by lacewings and certain other predators, the practical effectiveness of these products is largely undocumented.

Snakefly larvae are flattened and have a dark brown to shiny black head. They occur under bark and in leaf litter, where they feed on various soft-bodied insects such as wood borers and codling moth pupae. Snakeflies pupate without spinning a cocoon.

¾ INCH

LENGTH

Snakeflies, such as this adult *Raphidia* sp., have a shiny dark, flattened head and prolonged necklike thorax, giving them a serpentlike appearance. Adults are larger than lacewings, with clear, many-veined wings, a brown or dark reddish body, and an extended, tail-like ovipositor in females.

MANTIDS

Mantids (family Mantidae), commonly called praying mantids or mantises, are among the best-known insect predators. Mantids frequently occur naturally in gardens and are sold as egg cases for release.

Mantids are 2 to 4 inches (5–10 cm) long at maturity and are usually yellowish, green, or brown. They are easily recognized by their elongate thorax and

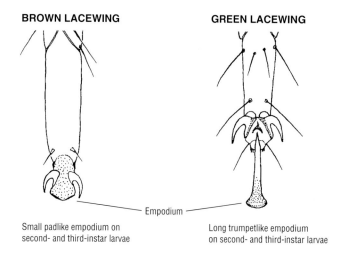

BROWN LACEWING

GREEN LACEWING

— Empodium —

Small padlike empodium on second- and third-instar larvae

Long trumpetlike empodium on second- and third-instar larvae

■ FIGURE 8-15. Mature green and brown lacewing larvae can be distinguished by the lobelike structure or empodium between the terminal claws on their feet. Second- and third-instar brown lacewings have a small padlike empodium; green lacewings and first-instar brown lacewings have a long trumpet-shaped empodium. Adapted from Peterson 1960, reprinted with permission from Helen H. Peterson.

grasping forelegs. Because mantids feed on many kinds of insects, including beneficial species, their beneficial value is limited. They often wait at flowers, where they prey on nectar- and pollen-feeding species, including honey bees, syrphid flies, and other beneficials. Mantids are also highly cannibalistic; they often eat their siblings as they hatch out of the egg case, and males are sometimes consumed by females after mating. Mantids are fascinating and make good study animals for anyone interested in insects; however, they are not recommended for use in controlling pests.

EARWIGS

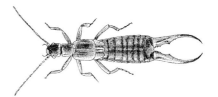

Earwigs (order Dermaptera) are common in gardens, landscapes, and many field and orchard crops and sometimes enter buildings. Although earwigs are pests when they feed on seedlings, blossoms, shoots, and soft fruits such as apricots, cherries, and peaches, they also help recycle organic matter by feeding on dead plant material. Most earwigs are beneficial predators that feed on other insects. Some, such as the striped earwig (*Labidura riparia*, Labiduridae) found principally in southern California, are mostly predaceous and only occasionally feed on plants.

At least seven species of earwigs occur in California; the most common is the European earwig (*Forficula auric-*

ularia, Forficulidae), which is distributed throughout much of the world. In apple orchards, the European earwig is a primary predator of apple aphid and woolly apple aphid. One study found that in comparison with lady beetles, European earwigs were better predators of woolly apple aphids (Asante 1995). Earwig populations have been augmented by rearing them on dog food and introducing them into apples and by providing earwigs with shelters of thin cardboard bands around trunks and straw scattered on orchard floors (Carroll and Hoyt 1984). Earwigs also

¾ INCH
LENGTH

This adult mantid fly (*Mantispilla* sp., Mantispidae) has a similar appearance and biology to snakeflies.

Mantids overwinter as eggs in hardened masses attached to bark; dozens of nymphs can emerge from a single egg case. The commonly sold egg cases of the Chinese mantid (*Tenodera aridifolia*) are thicker than those of the native species pictured here.

1⅕ INCHES
LENGTH

Mantids wait for prey with their legs upraised, creating the impression that they are praying. This immature (wingless) mantid is eating a house fly.

LENGTH ⊢

This predatory thrips nymph (*Franklinothrips vespiformis*) feeds on mites and pest thrips, including persea mite (*Oligonychus perseae*), avocado thrips (*Scirtothrips perseae*), and greenhouse thrips in avocado. Nymphs are recognized by their distinctive reddish orange abdomen. Adults are mostly blackish with a white band around their waist.

LENGTH ⊢

A larva of the black hunter thrips. Adults and larvae are entirely dark brown to black, sometimes with lighter-tipped antennae. Eggs and nymphs of various soft scales and mites are important prey.

LENGTH ⊢

This western flower thrips is a serious pest of some flower, fruit, and row crops. Although this species is a common pest, the yellowish larvae of western flower thrips feed on both plants and mite eggs. In certain crops, such as cotton, cucurbits, dry beans, figs, and strawberries, the value of western flower thrips as a mite predator normally outweighs the injury it causes to leaves.

LENGTH ⊢

This sixspotted thrips adult and its larvae are entirely predaceous, most commonly on mites. It is named for the three dark spots on each wing cover of the mostly pale-yellow adult. This predator can rapidly reduce high mite populations, but often not until after mites have become abundant and may have caused damage.

are beneficial predators of grain aphids (Sunderland and Vickerman 1980), cabbage looper pupae (Strandberg 1981), and eggs of velvetbean caterpillar (Buschman et al. 1977).

THRIPS

Thrips (order Thysanoptera) are tiny slender insects with long fringes on the margins of their wings. Although many species are plant feeders, some are important predators of mites and small, soft-bodied insects. Adult thrips are commonly yellowish or black. Larvae or nymphs are translucent white to yellowish. Thrips can produce many generations in a year and may move from crop to crop. Except for *Franklinothrips vespiformis*, the species pictured here occur throughout North America.

MITES

Although some mites feed on plants and can become pests, other mites are predators of pest mites, small insects, and insect eggs. Mites in the family Phytoseiidae are the best-known predators, but predaceous species occur in other mite families, including Anystidae, Laelapidae, and Stigmaeidae. Naturally occurring populations of predatory mites are extremely effective biological control agents in a range of crops and landscapes. Outbreaks of pest mites are frequently associated with the killing of predatory mites by broad-spectrum insecticides applied to control other pests, although plant water stress is another common cause of mite outbreaks. Augmentative release of predatory mites is recommended for use in certain crops (such as almonds or strawberries), some container-grown ornamentals, and greenhouse-grown vegetables.

Mites, unlike insects, do not have antennae, segmented bodies, or wings. Most mites pass through an egg stage, a six-legged larval stage, and two eight-legged immature (nymphal) stages before becoming adults (fig. 8-16). Most predaceous mites are long-legged and pear-shaped and are shinier than pest mites because they have fewer tiny hairs. Many are translucent, although after feeding they often take on the color of their host and may be bright white, yellow, red, or green. Predaceous mite eggs are more translucent, pearl-colored, and oblong than the eggs of plant-feeding mites, which are commonly spherical and colored or opaque.

To the naked eye, predator mites resemble moving specks. One way to distinguish plant-feeding mites from predaceous species is to closely observe infested foliage with a good hand lens. Predaceous species are much more active than plant-feeding pest species;

they stop moving only to feed. To make mites move, blow on them or touch them gently: predatory mites will move more quickly than pest mites. Under magnification, you can see that predatory mites have mouthparts that extend in front of their bodies to pierce prey; the mouthparts of pest mites extend downward to feed on plants. Because of their tiny size and diversity, most mites can be positively identified only by an expert.

In most of California and the southern United States, all stages of mites may be present year-round. In cold-winter areas, mites overwinter as adult females or as eggs on bark or in litter. At moderate temperatures, some species can complete a generation in 1 or 2 weeks.

Many predaceous mites feed not only on all stages of plant-feeding mites but also on insect eggs and immatures, such as scale crawlers and nymphs of thrips and whiteflies. Some predatory mites, especially *Amblyseius* and *Euseius* spp., are more generalized feeders and supple-

ment their arthropod diet with pollen, fungi, and leaf sap. *Phytoseiulus* spp. are very specific feeders, consuming only spider mites. Mites in some other genera (such as *Galendromus, Metaseiulus,* and *Neoseiulus*) have an intermediate diet, preferring spider mites but also feeding some on pollen and other food.

 Phytoseiulus persimilis feeding on a twospotted spider mite egg. *Phytoseiulus* is a very effective predator, each day eating several female spider mites or about 2 to 3 dozen spider mite eggs or immatures. Because it can eventually consume virtually all available prey, after which it disperses from plants, it must be reintroduced when spider mites reappear.

 Euseius tularensis preying on a citrus thrips nymph. It also feeds on citrus red mite and pollen and is active during the spring and fall flushes of new citrus growth. If citrus needs to be sprayed for thrips control, sabadilla, ryania, or abamectin are recommended to avoid severe mortality of thrips predators. Numbers of *E. tularensis* increase naturally if broad-spectrum insecticides such as carbamates and organophosphates are avoided.

 Western predatory mite feeding on a spider mite egg. This predator is about the size of the twospotted spider mite, but the predator lacks spots, ranges in color from cream to amber red, and is more pear-shaped than its prey. Its shiny, oval eggs are larger than spider mite eggs. The western predatory mite tolerates hot climates as long as the relative humidity is above about 50 percent. It is commonly found in almonds and grapes and is effectively released on many crops and ornamentals.

ENHANCING BIOLOGICAL CONTROL WHEN USING MITES

Avoiding the use of mite-disrupting pesticides and employing compatible cultural practices are two key methods of enhancing biological control of mites. Most carbamates, organophosphates, and pyrethroids can be extremely toxic to many predatory mites (table 6-2) and cause pest mite outbreaks when applied to control other pests (fig. 1-2). Because water stress and dusty conditions promote mite outbreaks, provide plants with appropriate irrigation and control excess dust from dirt roads near fields.

Prune the interior of citrus trees to increase predaceous mite populations in the exterior canopy, thereby reducing fruit-scarring by citrus thrips (Grafton-Cardwell and Ouyang 1995b).

In addition to the substantial pest control provided by conserving naturally occurring populations of predaceous

LENGTH

This *Hypoaspis* sp. (Laelapidae) is a predator of fungus gnats and other small soil-dwelling insects. Where present, this relatively large, reddish predator can be observed rapidly running on the surface of damp media.

Eggs of most predaceous mites are oblong, such as this egg of *Amblyseius cucumeris* shown next to the larger adult. Some *Amblyseius* species (*A. cucumeris, A. barkeri*) are released in greenhouses for control of thrips.

LENGTH

Anystis agilis (Anystidae) attacking a grape leafhopper nymph. This relatively large mite is also common in citrus, where it feeds on citrus thrips, psocids, and other small insects.

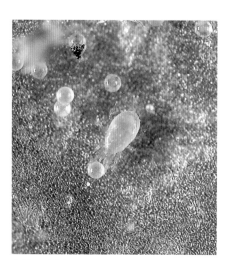

Neoseiulus or *Amblyseius californicus* preying on a spider mite egg. This predator is often released in greenhouses because it does well at moderate temperatures and requires a relative humidity of at least 65 percent. It can also be released outdoors where temperatures do not exceed about 85°F (30°C). Because *N. californicus* can survive on just pollen, it persists even when pest populations are low.

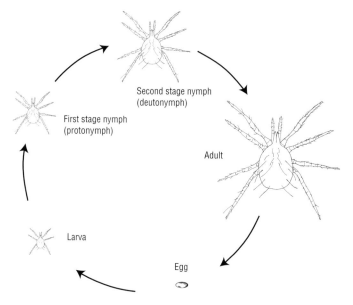

- FIGURE 8-16. Most mites, such as this *Phytoseiulus persimilis,* hatch from eggs and develop through three immature stages before becoming an adult. The adult female lays a light-orangish egg near prey that darkens before hatching. A six-legged mite larva emerges and in many predaceous species is relatively inactive, apparently not feeding before molting to the nymph stage. The eight-legged protonymph, deutonymph, and adult are active searchers and begin feeding almost immediately after molting. Adapted from Denmark and Schicha (1983).

TABLE 8-5. Commercially available predatory mites.

PREDATORY MITE	COMMENTS	REFERENCES
Specific predators of spider mites		
Phytoseiulus persimilis	An orangish predator active at 60 to 90% relative humidity (RH) and 70° to 80°F (21° to 27°C) or higher. A strain that tolerates temperatures up to about 100°F (38°C) is available. Feeds almost exclusively on spider mites; populations don't persist if spider mites are absent. Prefers and is often released against *Tetranychus* spp.	Easterbrook 1992; Osborne, Ehler, and Nechols 1985
Phytoseiulus longipes	Resembles *P. persimilis*, but will tolerate lower (40%) RH at 70°F (21°C). At higher temperatures it requires more humidity than *P. persimilis*. Feeds almost exclusively on spider mites.	Badii and McMurtry 1984
Prefer feeding on mites, but also consume pollen and other food		
Western predatory mite, *Metaseiulus (=Galendromus =Typhlodromus) occidentalis*	Somewhat shiny pear-shaped mite that takes on the color of recent food (beige to amber to red). Tolerates hot climates if RH is at least 50%. Some pesticide-tolerant strains are resistant to sulfur and certain carbamates and organophosphates. Feeds on mites from various genera and feeds also on pollen. Commonly released against *Tetranychus* spp. Less effective against European red mite because it cannot break that host's harder egg shell.	Badii and McMurtry 1984, Croft and MacRae 1992, Flaherty and Huffaker 1970, Hoy et al. 1984
Neoseiulus californicus	Translucent with colored spots. It tolerates temperatures up to 85° to 90°F (30° to 33°C), but needs RH of at least 65%. Commonly released in greenhouses. Feeds on mites in various genera and also on pollen. It persists well when pest populations are low.	
Generalized feeders on mites and insects, supplemented with pollen, fungi, and plant sap		
Amblyseius spp., *Amblyseius (=Neoseiulus) barkeri*	Feed on various mites and small insects. Persist on pollen and can survive when prey populations are low. Released to control broad mite and flower thrips in greenhouses.	Fan and Petitt 1994, van Houten et al. 1995
Euseius spp.	Shiny white, yellow, or reddish mites that take on the color of their food. Avoid light and will run quickly across the leaf if held in bright sunlight. Feed on most mites and various small insects, including citrus red mite and citrus thrips. Feed on various pollens, honeydew, and leaf sap, and can survive when prey are scarce.	Grafton-Cardwell, Eller, and O'Connell 1995; Grafton-Cardwell and Ouyang 1995a, 1995b; Jones and Morse 1995; Ouyang, Grafton-Cardwell, and Bugg 1992
Hypoaspis (=Geolaelaps) miles	Reddish, relatively large, soil-dwelling mite that readily disperses across the surface of moist media. Adults feed on thrips pupae, fungus gnat larvae, and other small invertebrates near the surface of media. Can survive 3 to 4 weeks without insect prey.	Wright and Chambers 1994

Sources: Grafton-Cardwell and Ouyang 1995a, McMurtry and Croft 1997.

mites, *Phytoseiulus persimilis, Neoseiulus californicus,* and the western predatory mite (*Galendromus* or *Metaseiulus occidentalis*) are among the species that can be purchased and released to control Pacific spider mite, twospotted mite, and some other pests if resident predators are insufficient (table 8-5). The western predatory mite, and possibly other available mites, are resistant to sulfur and certain carbamates and organophosphates and can be used in treated orchards (Hoy 1984, Roush and Hoy 1981).

The best-documented effective release of predaceous mites is in greenhouses (Osborne and Ehler 1981; Osborne, Ehler, and Nechols 1985) to control spider mites infesting vegetables such as tomatoes, cucumbers, and peppers (Fan and Petitt 1994, Hussey and Scopes 1985, van Lenteren and Woets 1988) and ornamentals such as roses (Field and Hoy 1984, Zhang and Sanderson 1995). Field releases are recommended in almonds (Hoy 1984, Hoy et al. 1984) and strawberries (Trumble and Morse 1993, Pickel et al. 1996) and are increasingly being used in other crops, including corn, cotton, and grapes.

Specific research-based guidelines for mite release have not been established for most situations. A general augmentation strategy is to begin making inoculative releases when there are sufficient numbers of spider mites or other hosts to feed predators and encourage their reproduction, but before pest populations rise to high levels. Successful inoculation over a period of several months to 2 years may permanently establish an effective predator population in some perennial crops (such as almonds) or in natural areas bordering crops (such as strawberries and beans) (McMurtry et al. 1978). These border areas can serve as a reservoir for predaceous mites that will then colonize the crops. In annual crops, repeated releases each growing season may be necessary (see the discussions of using predatory mite nurse plants

A field worker applying predatory mites to strawberries. These predators have been commercially reared and shipped on mite-infested bean plants, which are distributed at intervals throughout the crop. Some insectaries ship mites in containers mixed with packing material, such as vermiculite, that can be sprinkled onto plants. In addition to the broad-spectrum insecticides that are toxic to mites, benomyl fungicides applied to strawberries can prevent *P. persimilis* from reproducing.

with strawberries and releasing natural enemies effectively in chapter 6).

SPIDERS

Spiders (order Araneae, Class Arachnida) are very common invertebrates. Unlike insects, which have six legs and three main body parts, spiders have eight legs and two main body parts (fig. 8-17). Spiders are classified in the arachnid group along with mites.

All spiders are predaceous; they eat mainly insects, other spiders, and related arthropods. Some species capture prey in webs and others stalk insects across the ground or vegetation and pounce on

Hololena nedra is a mostly brownish funnel weaver spider (family Agelenidae) with a dark gray band on each side of its abdomen. Its prey includes beetles, bugs, flies, and moths; leafhoppers are a primary prey in San Joaquin Valley vineyards.

LENGTH

them. About 50 families of spiders occur in the United States. The family of many common spiders can be identified by observing spiders' body shape, eye pattern, type of web, and hunting or other behavior in the field (table 8-6).

Although naturally occurring spiders are believed to significantly reduce insect pests in some situations (Provencher and Riechert 1994, Riechert and Bishop 1990), there is little information on how to manipulate spiders to improve pest control, and none are commercially available for augmentative releases. Some species are relatively indiscriminate feeders, preying on a variety of insects. Other species are more specific in their prey and habitat preferences. For example, because they mostly wait for prey in aerial webs, orb weavers are more likely to capture adult flying insects than immature or ground-dwelling insects.

Spiders seek to avoid people, and most are harmless to humans. The black widow and less common brown spiders or recluse spiders (*Loxosceles* spp., commonly called brown recluses), can cause painful injuries or systemic illness, but they seldom if ever are fatal to healthy adults (Barr, Hickman, and Koehler 1984). In comparison with

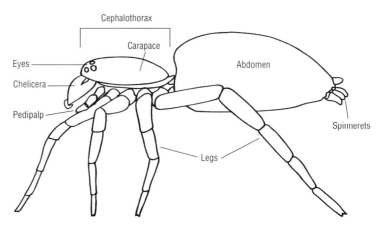

■ FIGURE 8-17. Side view of a spider body. Spiders have two main body parts: an abdomen and a one-piece head and thorax (cephalothorax). The cephalothorax contains 4 pairs of legs, eyes (usually 8), mouthparts (chelicerae), and leglike sensory organs (pedipalps). The shieldlike top of the cephalothorax is called a carapace. Rear spinnerets on the abdomen produce silk. Adapted from Costello et al. (1995).

LENGTH Cobweb weavers (Theridiidae), such as this female *Theridion melanurum*, snare prey in sticky, irregularly spun, clumpy silk.

LENGTH *Theridion dilutum* is a diurnal cobweb weaver that waits for prey on the underside of leaves. This predator is common in California vineyards, where its prey includes flies, leafhoppers, and mites.

¾ INCH LENGTH This jumping spider (*Phidippus* sp., Salticidae) is eating a house fly.

LENGTH The agrarian sac spider (*Cheiracanthium inclusum*, Clubionidae) has a pale yellow to greenish body with a dark brown to black carapace and dark legs. This is the most common spider in San Joaquin Valley vineyards, where its prey includes leafhoppers, various flies, and other spiders. Some related sac spiders, such as *Trachelas pacificus*, are sometimes placed in the ant mimic spider group, family Corinnidae.

LENGTH This house spider (*Tegeneria* sp., Agelenidae) is commonly observed on walls and ceilings in homes and also lives outdoors. In homes, it feeds on insects and other spiders that find their way inside.

LENGTH This flower spider or goldenrod crab spider (*Misumena vatia*, Thomisidae) waits on blossoms for bees, flies, and other prey that visit flowers. Some species of flower spiders slowly change color to blend in with their background.

TABLE 8-6. Common spider families.

Agelenidae, funnel weavers or grass spiders

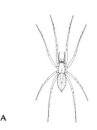

A

Funnel weavers are sit-and-wait predators that feed during the day and night on the ground and in most types of vegetation, including low-growing plants and trees. They spin funnel-shaped webs, often with a several-inch-wide, flat extension covering plants or soil. The spider waits in the hole of its web. When it detects vibrations from an insect that flew or walked into the web, the spider runs out, captures and bites the prey, then carries it back into the funnel to be eaten. Webs on low vegetation become conspicuous in morning light after collecting dew. Funnel weavers have six or eight eyes, all about the same size, arranged in two rows. About 300 species in North America.

Araneidae, orb weavers or garden spiders

B

Orb weavers feed on insects that fly, fall, or are blown into their web. Their elaborate silken webs are spun in concentric circles. Spiderlings often make symmetrical webs; mature spiders may spin a more specialized design that is helpful in identifying certain species. The spider rests at the center of its web or hides in a shelter near the edge, waiting for prey to become entangled. Orb weavers generally have poor vision and rely on web vibrations to locate and identify prey. About 3,500 species occur in the world, with about 200 in America north of Mexico.

Clubionidae (including Corinnidae), sac spiders or twoclawed hunting spiders

C

These spiders stalk and capture prey that is walking or resting on surfaces. They spin silken tubes or sacs under bark, among leaves, and in low plants or on the ground, where they hide during the day or retreat after hunting. Sac spiders commonly are nocturnal, medium-sized, pale spiders with few markings. About 200 species in North America.

Linyphiidae (=Microphantidae), dwarf spiders

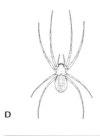

D

Dwarf spiders prey on insects that fall, walk, or land in their web. They are diurnal (day-active) spiders occurring in the plant canopy and among litter on the ground. They produce sheetlike webs on the surface of plants or soil and are common in some field and vegetable crops. Most are relatively small. Several hundred species in North America.

Lycosidae, wolf spiders

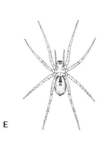

E

Eyes

Wolf spiders prey on insects that are walking or resting on the ground. They actively hunt in the open during the day and night, and are often observed on the ground in litter and on low vegetation. They also occur in burrows and under debris on soil. Instead of spinning webs to catch prey, they make a small, thick web in which they rest. Wolf spiders have a distinctive pattern of eyes: four small eyes in front in a straight row, one middle pair of larger eyes, and one rear pair of widely spaced eyes on top of the head. They have long hairy legs. They are usually black and white or strongly contrasting light and dark, which can make them difficult to discern unless they are moving. About 200 species in North America.

Oxyopidae, lynx spiders

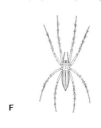

F

Eyes

Lynx spiders stalk and capture resting or walking insects. They are active hunters with good vision. Most have spiny legs and a brightly colored body that tapers sharply toward the rear. Lynx spiders have four pairs of eyes grouped in a hexagon. About 2 dozen known species in North America; many more in tropical areas.

TABLE 8-6. Common spider families *(continued)*.

Salticidae, jumping spiders

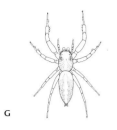

G

Eyes

Jumping spiders are day-active hunters in plants or on the ground. They make no web; instead they stalk and pounce on prey by jumping distances many times their body length. Jumping spiders have a distinctive pattern of eyes in three rows: the first row of four eyes, with large and distinctive middle eyes; a second row of two very small to minute eyes; and a third row of two medium-sized eyes. They usually have an iridescent, metallic-colored abdomen and black carapace. More than 5,000 species worldwide, about 300 in America north of Mexico.

Thomisidae, crab spiders or flower spiders

I

Eyes

Crab spiders stalk and capture insects walking or resting on surfaces. They are diurnal hunters that do not spin webs. Their front two pairs of legs are enlarged and extend beyond the side of their flattened body, making them look like tiny crabs. Their small eyes occur in two slightly curved rows, with the top row often much wider than the lower row. Over 200 species in North America.

Theridiidae, cobweb, cobweb weaver, or combfooted spiders

H

Cobweb spiders feed on insects that walk or fly into their webs. These spiders are almost always found hanging upside down by their claws in irregularly spun, sticky webs, waiting for prey. The spider is usually concealed in a corner of the web, in a silken tent, or behind debris. This group includes the black widow spider, which produces relatively thick silk that feels rough and sticky. Cobweb spiders generally have a soft, round, bulbous abdomen and slender legs without spines. Over 200 species in America north of Mexico.

Sources: Costello et al. 1995, Kaston 1978, Levi and Levi 1990 (A–G *from* Davies 1986, reprinted with permission from Queensland Museum; H *from* Barker et al. 1978; I *from* Emerton and Frost 1961, reprinted with permission from Dover Publications).

black widows or recluse spiders, more people are bitten by the common house-dwelling agrarian sac or yellow sac spiders, and these bites cause relatively mild symptoms.

BLACK WIDOW

The adult female black widow spider (*Latrodectus mactans* =*L. hesperus*, family Theridiidae) has a black shiny body, slender legs, and a red hourglass-shaped mark on the underside of her large round abdomen. It occurs indoors and outdoors in fields and vineyards. Although its bite can be serious, black widows tend not to bite people unless you actually touch the spider or disturb its web, especially if egg sacs are present. Black widow spiders prefer dark places, where they wait in webs for

LENGTH

Female wolf spiders (Lycosidae) such as this *Lycosa* sp. generally carry their young spiderlings atop their abdomens.

LENGTH

A female black widow suspended upside down in her web. Above is her egg sac, which is roundish to pear-shaped and about ½ inch (12 mm) in diameter. It consists of tough, pale-colored silk and contains up to several hundred eggs. The tiny spiderlings that emerge will balloon to another location by producing silken thread and floating away with the wind.

This black and yellow garden spider or yellow garden argiope (*Argiope aurantia*, Araneidae) is often observed suspended in its web in gardens and landscapes. It eats almost any flying insect that becomes caught in its web.

This adult redbanded crab spider (*Misumenoides formosipes*, Thomisidae) crouches in wait. It ambushes prey that visit flowers, such as small bees, bugs, and flies.

This tiny dwarf spider (*Erigone* sp., Linyphiidae =Microphantidae) preys on small insects and mites in orchards, strawberries, and vineyards. It makes irregular crisscross or sheetlike webs on foliage.

prey, which include flying insects such as flies and ground dwellers such as crickets and cockroaches.

The harmless male black widow varies from light greenish gray to dark brown or blackish and has cream-colored patches and a light to brownish lengthwise band. This male coloring resembles young female black widows. The hourglass marking on the underside of both immature females and males is yellow to orangish.

SNAILS AND SLUGS

Snails (phylum Mollusca) are soft-bodied, unsegmented invertebrates with a hard shell. Although most snails feed on plants or decaying organic matter, some species are predaceous. Snails and slugs have many natural enemies, including predatory beetles (such as Cantharidae, Carabidae, Silphidae), flies (principally Phoridae and Sciomyzidae), and pathogens. Amphibians, birds, fish (on aquatic mollusks), insectivorous mammals, and reptiles also prey on mollusks. Toads were formerly kept in greenhouses and sometimes still are to consume slugs (Godan 1983).

DECOLLATE SNAIL

The commercially available decollate snail was introduced into Southern California from the Mediterranean in the 1970s to control the brown garden snail. The decollate snail is now an important brown garden snail predator in citrus (Anonymous 1991a; Sakovich, Bailey, and Fisher 1984). Each snail lays about 500 eggs during its lifetime. The snails live in litter on soil, emerging to feed when it is dark and damp. Although decollate snails feed mostly on other snails and decaying organic matter, research indicates they may help control citrus thrips by feeding on thrips pupae in soil (Schweizer and Morse 1997). Decollates also eat young seedlings, so it may not be desirable to release them in gardens.

The best time to introduce and establish decollate snails in California is during warm, damp weather from February through May. Establishment and proper maintenance of decollate snail populations can permanently reduce brown garden snail populations in citrus to insignificant levels in 4 to 10 years (Grafton-Cardwell et al. 1996). Decollate snail introductions in California are permitted only in certain San Joaquin Valley and southern counties. Releases in other areas are illegal because they might decimate native snail and slug populations of ecological importance in natural areas. Check with your county agricultural commissioner or local wildlife protection agency to determine whether decollate snail introductions are permitted in your area.

| 1 INCH |
| LENGTH |

The predatory decollate snail's shell is elongate, tapered, and about 1 inch (25 mm) long. As it grows, the shell's tip becomes brittle and often becomes irregularly broken off at the tip.

| 5 INCHES |
| LENGTH |

About 2 dozen bat species occur in Pacific Coast states; most feed exclusively on insects. Some people encourage bats by providing nesting sites outside of inhabited buildings (Long 1996). Bats, such as these *Myotis* sp. evening bats, can be prevented from living in the same buildings as people by installing barriers that allow bats only to exit or by plugging their access holes after dark while bats are outside feeding.

BIRDS AND OTHER VERTEBRATES

Insects are important food for many birds, mammals, reptiles, and amphibians. Because bats and many rodents, snakes, and frogs feed largely on insects, they are beneficial, even though some people may dislike these often strange and secretive vertebrates. Many species of birds feed almost exclusively on insects and other species that normally feed on seeds or plants often rely on insects to feed their nestlings. Caterpillars and other larvae are apparently the insects most commonly fed upon by many birds (fig. 8-18). Muscovy ducks (*Cairina moschata*) can be very effective in controlling house fly maggots in enclosed dairy and swine facilities (Glofcheskie and Gordon 1993). Domesticated chickens (*Gallus gallus*) have long been used for fly maggot control.

Birds are believed to help keep insect populations low in many situations. The importance of insect-eating birds has probably been best studied in forests and woodlands (Dahlsten et al. 1990, Kleintjes and Dahlsten 1994). In addition to eating insects, birds may also help to limit pest populations by spreading insect viruses in bird feces (Entwistle et al. 1993, Smirnoff 1959). Some birds help to control pests other than insects by, for example, eating weed seeds or rodents.

Populations of desirable birds can be increased by growing a mixture of trees, shrubs, and ground covers of various sizes, species, and densities and by providing water and supplemental food (Tilghman 1987). Many insect-eating birds nest in cavities in dead trees. Dead and dying trees can rarely be left where people live and work because they are hazardous and may fall; a practical alternative is to provide nesting boxes or bird houses. The value of dead and dying trees as wildlife habitat should be assessed along with the hazards when considering tree removal.

Cliff swallows (*Hirundo pyrrhonota*, Hirundinidae) built these nests on a rock face. These birds feed almost exclusively on insects, which they catch in flight. Cliff swallows are protected by law because they are beneficial predators. Photo by Richard Eng.

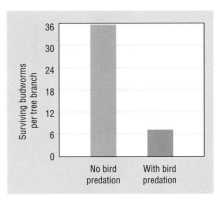

■ FIGURE 8-18. Birds are important predators of budworms and other conifer pests (Crawford and Jennings 1989). Six times more budworms survived on fir trees caged to exclude birds than on fir trees with bird predation (Torgersen et al. 1984, Torgersen and Torgersen 1995).

CHAPTER NINE

PATHOGENS OF ARTHROPODS

Magnifying glass signifies that subject is less than 1 mm.

PATHOGENS ARE infectious microorganisms that injure or kill their hosts. Some pathogens cause disease in desirable plants or animals. However, many other species of microorganisms are beneficial: they degrade toxins, build soils, and produce plant nutrients. Beneficial pathogens are important in controlling certain weeds (see chapter 5), plant-damaging pathogens (chapter 3), and vertebrates. Pathogens that kill insects and other invertebrates are discussed here.

Bacteria, fungi, nematodes, protozoans, and viruses are the most common groups of pathogens. Although some are commercially available as biological or microbial insecticides (table 9-1), many occur naturally in the field and can rapidly decimate pest populations when conditions are right. Except for nematodes, pathogens used for pest control must be registered and labeled in accordance with regulations that also apply to chemical pesticides. Therefore, the testing requirements and information on microbial insecticides can be more extensive than for commercially available predators and parasites, which are largely unregulated.

Microbial insecticides are essentially nontoxic to people and other vertebrates. Many microbials specifically attack certain pests and do not affect nontarget beneficial insects. Microbials generally break down rapidly in the environment, reducing the likelihood that pesticide resistance will develop. However, because they are organisms, microbial pesticides need careful handling; for example, storing them under very hot or freezing conditions can reduce their viability or kill them. Their lack of persistence also means that repeated applications may be necessary to provide adequate control in many situations. Thorough coverage and proper application timing are critical to effective use of most microbial pesticides. Because most are pest-specific and some are difficult to produce, microbial insecticides may have limited availability or be expensive.

■

BACTERIA

Bacteria are microscopic single-celled organisms that feed in or on organic matter or living organisms. Although many bacteria must remain in contact with a host or organic debris to survive, certain bacteria that infect insects can persist outside hosts and spread by forming spores. Bacteria also disperse in water or on infested insects, plants, soil, or equipment.

TABLE 9-1. Pathogens commercially available as biological or microbial insecticides.

PATHOGEN NAME	TYPE	PESTS CONTROLLED
Autographa californica NPV[†]	V	alfalfa looper larvae
Bacillus lentimorbus[†]	B	Japanese beetle larvae, turf grubs, white grubs
Bacillus popilliae[†]	B	Japanese beetle larvae, turf grubs, white grubs
Bacillus sphaericus[†]	B	mosquito larvae
Bacillus thuringiensis ssp. *aizawai*	B	greater wax moth caterpillars in beehives
Bacillus thuringiensis ssp. *kurstaki*	B	caterpillars of butterflies and moths
Bacillus thuringiensis ssp. *israelensis*	B	mosquito, black fly, and fungus gnat larvae
Bacillus thuringiensis ssp. *san diego* or *tenebrionis*	B	leaf beetle larvae, including Colorado potato beetle and elm leaf beetle
Beauveria bassiana[†]	F	aphids, crickets, grasshoppers, locusts, thrips, whiteflies, and some other exposed insects, especially under humid environments
beet armyworm NPV[†]	V	beet armyworm larvae
codling moth granulosis virus	V	codling moth larvae
Douglas fir tussock moth NPV[†]	V	Douglas fir tussock moth larvae
gypsy moth NPV	V	gypsy moth larvae
Heliothis NPV[†]	V	bollworm, tobacco budworm
Heterorhabditis bacteriophora	N	flea beetles, Japanese beetle larvae, root maggots, turf grubs, weevils, white grubs, and certain other soil-dwelling insects
Hirsutella thompsonii[†]	F	citrus rust mite
Lagenidium giganteum	F	mosquito larvae
Metarhizium anisopliae[†]	F	cockroaches, flies
Nosema locustae[†]	P	grasshoppers, crickets, locusts
Paecilomyces fumosoroseus[†]	F	aphids, whiteflies, and some other exposed insects, especially under humid environments
sawfly nuclear polyhedrosis virus[†]	V	pine sawfly larvae
Steinernema carpocapsae	N	carpenterworm, clearwing moth larvae, flea beetles, Japanese beetle larvae, root maggots, turf grubs, weevils, white grubs, and certain other soil-dwelling insects
Steinernema feltiae	N	fungus gnat larvae, certain other soil-dwelling insects

KEY
B   bacterium
F   fungus
N   nematode
P   protozoan
V   virus

† Product availability may be limited, and materials may not be registered in California.

Note: Commercial products are available under various trade names; contact suppliers or check labels for specific products and registered uses. Watch for availability of new products. See table 6-3 for commercially available predators and parasites to control these pests and some other groups. Consult the index for scientific names of the invertebrates mentioned here.

Sources: Cook et al. 1996; Quarles 1996b; Starnes, Liu, and Marrone 1993.

Over 100 species of bacteria have been identified as insect pathogens, but only certain *Bacillus* spp. are available commercially. *Bacillus thuringiensis* (Bt) is the most widely used pathogen for pest control (Starnes, Liu, and Marrone 1993; Trumble, Carson, and White 1994). Various Bt subspecies (formerly called *strains*) are commercially available for controlling foliage-feeding caterpillars, leaf beetle larvae, and larvae of mosquitoes, black flies, and fungus gnats (table 9-1). Because in most groups, only younger-stage insects are highly susceptible to Bt, it must be applied when early instars predominate. Exceptions are hornworms (*Manduca* spp., Sphingidae) and some caterpillars

A healthy green larva and two black second-instar elm leaf beetles killed by *Bacillus thuringiensis* ssp. *tenebrionis* (Btt). This microbial pesticide is important for controlling Colorado potato beetle, which has developed resistance to many insecticides. As a result of its widespread, repeated application, some populations of Colorado potato beetle and certain other species, including diamondback moth and stored grain pests, have developed resistance to Bt.

A healthy larva and two brownish Indian meal moth larvae killed by *Bacillus thuringiensis*, which is most commonly applied against foliage-feeding caterpillars. Because Bt is non-toxic to humans, it also can be applied to control lepidopteran pests of stored walnuts and in other situations where residues of broad-spectrum pesticides might be unacceptable.

in families Geometridae and Pieridae; later instars of these are quite susceptible to Bt.

As with most bacteria, Bt must be eaten by the insect before it can infect and kill its host. Once consumed, a Bt protein crystal (endotoxin) destroys the insect's gut, allowing bacterial spores to pass from the insect's stomach into its blood system, where the bacteria reproduce. Proper application timing (when susceptible stages are feeding) and good spray coverage are critical for effective control. Bacteria-infected insects become lethargic, stop feeding, and excrete liquid. After death, the insect's body becomes dark and soft and eventually decomposes into a dark, liquidy, putrid mass (fig. 9-1, table 9-2).

■ NEMATODES

Nematodes are tiny (usually microscopic) roundworms. Many species are free-living in soil or water and feed on bacteria and fungi. Pest species feed on plants or parasitize people and animals. Some nematodes are beneficial because they kill pest nematodes or insects; for example, nematodes in the genus *Mononchus* feed on other nematodes.

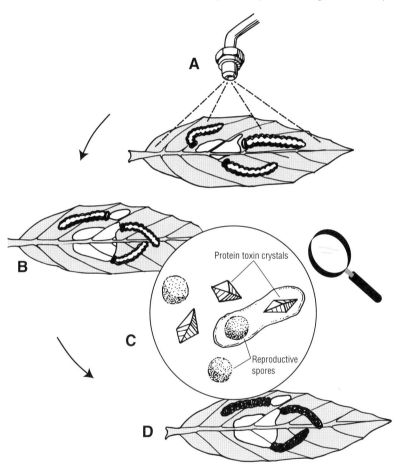

■ FIGURE 9-1. Several subspecies of *Bacillus thuringiensis* (Bt) are widely used to control certain insects. The Bt subspecies *kurstaki* controls moth and butterfly caterpillars: A: Bt must be sprayed during warm, dry conditions to thoroughly cover foliage where young caterpillars are actively feeding. B: Within about one day of consuming treated foliage, caterpillars become infected, relatively inactive, and stop feeding. C: An enlarged view of Bt in the gut of a caterpillar. The natural bacteria are rod-shaped and contain reproductive spores and protein toxin crystals (endotoxins). The spores and protein crystals are separate components in some commercial Bt formulations, and these separate components and one whole bacterium (greatly enlarged) are shown here. D: Within several days of ingesting Bt, caterpillars darken and die, and their carcasses eventually decompose into a dark, liquidy, putrid mass.

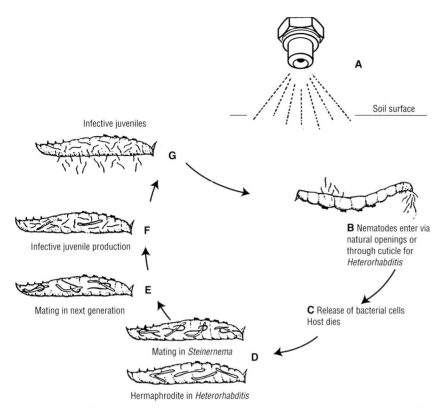

Infective juveniles

Nematodes enter via natural openings or through cuticle for *Heterorhabditis*

Release of bacterial cells Host dies

Infective juvenile production

Mating in next generation

Mating in *Steinernema*

Hermaphrodite in *Heterorhabditis*

Soil surface

■ FIGURE 9-2. Life cycle of beneficial nematodes: A: Infective-stage nematodes are applied to soil. B: The nematodes seek a host and enter it. C: Once inside, the host is killed by nematodes and mutualistic bacteria carried by nematodes. D: Nematodes feed, grow, mature, and reproduce. In *Heterorhabditis* spp., the infective-stage nematodes that originally entered the host mature into reproductive individuals that contain both male and female organs (hermaphrodites); there are no separate male and female nematodes in this first generation of *Heterorhabditis*. E: In both genera, the next generation of nematodes within the host produce both males and females that mate. F: Females then produce infective-stage juvenile nematodes inside the dead host. G: Infective nematodes exit and seek hosts. Nematodes persist in media and in dead hosts and, under suitable conditions, can provide residual control. The entire life cycle from infection of the host to release of the new infective generation takes 7 to 14 days. Adapted from Kaya 1993.

Nematodes that kill insects are discussed here. These beneficials are termed *entomopathogenic* nematodes because hosts are killed within several days of infection by the nematode in combination with associated bacteria (fig. 9-2).

Many insects are susceptible to entomopathogenic nematodes, such as *Heterorhabditis* and *Steinernema* spp. (families Heterorhabditidae and Steinernematidae, respectively). Commercially available nematodes are applied primarily against soil-dwelling insects (Hom 1994) including larvae and pupae of fungus gnats (Harris, Oetting, and Gardner 1995), white grubs and other turf pests (Cowles and Kido 1996), and weevils (Burlando, Kaya, and Timper

1993; Kaya and Gaugler 1993). These nematodes can infect many kinds of insects living in moist or humid environments, such as some ground-dwelling adult pests hiding in litter and larvae of armyworms, cutworms, and wireworms (Choo, Koppenhöfer, and Kaya 1996; Kaya 1993). Certain species that bore into plant tissue, such as clearwing moth larvae (Gill, Davidson, and Raupp 1992; Gill et al. 1994) and leafminers (Hara et al. 1993), can be controlled by nematodes if nematodes can be effectively applied to reach their hosts. At high relative humidity, certain nematode species are more tolerant of desiccation (such as *Steinernema carpocapsae*) and may be able to control

foliage-feeding insects such as the diamondback moth (*Plutella xylostella*) (Baur, Kaya, and Thurston 1995).

Heterorhabditis bacteriophora (formerly *H. heliothidis*) and *Steinernema carpocapsae* are the species most commonly available commercially. As with all natural enemies, these entomopathogenic nematode species differ in the conditions under which they are most effective (fig. 9-3). Drenching soil with entomopathogenic nematodes when pest larvae and pupae are present can infect and kill the immature insects if the soil is at least 60°F (16°C) and moist, but not soggy before application and for 2 weeks afterwards. Foliar sprays may control leafminers (*Liriomyza* spp.) if foliage surfaces are wet for several hours after application (as is common in greenhouses) and applications are made after dark (because nematodes can be killed by solar radiation). A squeeze bottle applicator with the nozzle inserted into tunnel openings or a hand sprayer used to thoroughly drench bark may provide control of carpenterworms and clearwing moth larvae infesting tree trunks and limbs.

■

FUNGI

Fungi are multicellular organisms usually composed of fine, threadlike structures (hyphae). These hyphae form a network or mass (mycelium) that grow on or through the host. Most insect-pathogenic fungi spread by producing minute, seedlike spores (conidia). Insect-pathogenic fungal spores are dispersed in water, soil, wind, or on insects, and some plant-pathogenic fungi are spread on contaminated equipment and people.

Virtually all groups of insects are susceptible to certain fungal diseases, which can cause widespread infections that rapidly decimate insect populations (Hajek and Leger 1994). Once in contact with the insect's body, conidia germinate, penetrate the cuticle, and infect the insect. Unlike bacteria and viruses, fungi do not have to be eaten

A healthy, soil-dwelling scarab beetle grub (*Cyclocephala hirta*, photo left) is whitish. When killed by *Heterorhabditis* spp. nematodes, the host insect often turns reddish and its internal contents may become sticky or moist. The red color of the grub killed by *Heterorhabditis bacteriophora* is due to a pigment produced by the mutualistic bacterium *Photorhabdus luminescens*, which is associated with the entomopathogenic nematode.

LENGTH Insects killed by *Steinernema* spp. turn yellowish to brown. Shown here are a healthy black vine weevil pupa and larva (left) and two yellowish pupae and a brownish larva infected with *Steinernema feltiae* nematodes (right).

LENGTH The fuzzy orangish rose aphids seen here have been killed by a naturally occurring fungal disease that develops under moist conditions.

LENGTH *Beauveria bassiana* killed the rice water weevil shown here at the bottom and covered its body with whitish fungal mycelia.

to infect hosts. Fungi are the only pathogens important for controlling Homoptera, because Homoptera suck plant juices, thereby avoiding ingestion of contaminated plant surfaces that lead to pathogen infection in chewing insects. Naturally occurring fungi usually require humid conditions to cause an epidemic in which they spread from one insect to another. However, good spray coverage and repeated applications may allow some commercial fungal applications to be effective when humidity is relatively low.

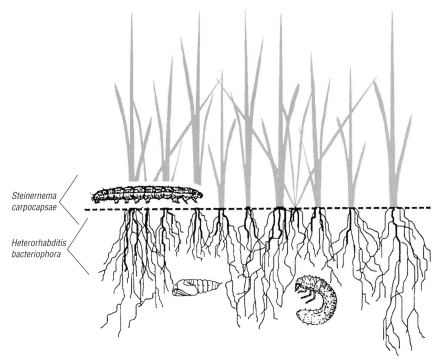

Steinernema carpocapsae

Heterorhabditis bacteriophora

■ FIGURE 9-3. All natural enemies differ in the conditions under which they are effective. Understanding the biology greatly increases the effectiveness of control actions. As shown here, a combination of two commercially available entomopathogenic nematode species can be applied to control different life stages of the same insect pest or more than one species occupying different habitats. For example, *Steinernema carpocapsae* is a sit-and-wait forager that infects hosts that move near the soil surface, such as cutworms and armyworms. *Heterorhabditis bacteriophora* actively searches for prey below the soil surface; it attacks hosts such as root-feeding grubs and underground pupae of moths (Choo, Koppenhöfer, and Kaya 1996; Kaya 1993). As with virtually all natural enemies, besides differences in behavior and habitat, nematode species vary in effectiveness depending on host species (Caroli, Glazer, and Gaugler 1996). *Source:* Kaya 1993.

TABLE 9-2. Characteristics of caterpillars killed by virus (NPV), fungus, and bacteria (*Bacillus thuringiensis* or Bt). Although caterpillars killed by nuclear polyhedrosis virus (NPV), fungus (*Entomophaga maimaiga*), or *Bacillus thuringiensis* (Bt) can sometimes be distinguished in the field, there is considerable variation in cadaver symptoms, which are illustrated here with gypsy moth larvae. Microscopic examination is necessary to distinguish precisely the cause of caterpillar death. Microscopic size, difficulties in handling them, and confusion among symptoms are among the reasons why many beneficial pathogens have been overlooked and have been poorly studied.

CATERPILLARS KILLED BY:

CHARACTERISTICS	NPV	FUNGUS	BT
typical appearance of dead larvae	abdominal prolegs / thoracic legs	abdominal prolegs / larval hair	mostly younger larvae stop feeding or become paralyzed, darken, and slowly die; dark liquid may be excreted from the mouth and anus
occurrence	widespread only when host larvae are very abundant; affects mostly older larvae, just before they pupate	especially prevalent under prolonged, rainy or humid conditions, even if larval populations are not high	within several days after spraying Bt, infection does not naturally occur; affects mostly young larvae
body position	frequently hangs as an inverted V with both front and back ends down, attached by only prolegs at the body center; may also hang straight down	often attached vertically with head down	usually not attached to surface of plant or bark
body shape	often swollen behind the downward-hanging head, where liquified contents accumulate	not swollen behind the head, although may be stretched thin	typically soft and shapeless, but posture may look relatively normal
abdominal prolegs	more or less paired together; if projecting are on one side of body	each proleg in a pair often extends stiffly at 90° angle on opposite sides of the body	appendages held normally or adhere to moist body, giving somewhat shapeless appearance
body texture and contents	soft, moist, filled with brown liquid	dry and stiff; may have tiny white spores on surface	dark and soft, but less liquidy than with NPV
cadaver persistence	usually degenerates rapidly	may remain on trunks or bark through winter	often dries and shrivels, but integument remains intact

Sources: Hajek 1994, Hajek and Roberts 1992, Hajek and Snyder 1992, in part adapted from illustration by Tana L. Ebaugh.

Although over 700 species of fungi are known to be associated with insects, relatively little is known about how much pest control is naturally exerted by fungi. For example, beginning in 1989, a fungus (*Entomophaga maimaiga*) of uncertain origin began killing many gypsy moth larvae in the northeastern United States (Hajek, Humber, and Elkinton 1995). This pathogen may have been present for many years and was overlooked because caterpillars killed by the fungus can be confused with those killed by gypsy moth nuclear polyhedrosis virus (table 9-2).

In addition to naturally occurring species, certain fungi can be purchased and applied to control pests. *Verticillium lecanii* has been used for insect pest control in greenhouses in Europe for many years. Fungi such as *Beauveria bassiana*, *Paecilomyces* spp., and *V. lecanii* are of commercial interest because they infect many different kinds of insects, and the potential market for broad-spectrum microbial pesticides is relatively large. These fungi are being produced and applied experimentally in the United States; some have recently become commercially available (table 9-1).

Insects infected with fungi become lethargic, stop feeding, and may move to the top or edge of plants. Their dead bodies may become rubbery, turn into empty shells, or appear swollen. Pale to dark flocculent hyphal strands may grow out of the body or cover the entire dead insect.

VIRUSES

Viruses are submicroscopic particles that infect living cells and alter the host's development. Viruses require a living host in which to reproduce and generally do not survive very long outside live organisms. Many plant-damaging viruses are spread by plant-feeding insects, especially aphids and thrips. Natural outbreaks of most insect-pathogenic viruses are difficult to predict and do not occur until pests are abundant.

Baculoviruses are arthropod-specific pathogens that are harmless to plants, mammals, and nontarget insects. Some of these beneficial baculoviruses are spread by birds (Entwistle et al. 1993, Smirnoff 1959), parasites (Heinz et al.

1⅜ INCHES
LENGTH

Larvae killed by fungi or viruses often hang limply from plants, as with these two cadavers. A healthy silverspotted tiger moth caterpillar is on top of this Monterey pine twig. Such pathogens (especially viruses) are commonly called caterpillar wilt disease. The infected insect bodies eventually produce more pathogen particles that contaminate plants and infect other caterpillars that contact or eat those plant parts.

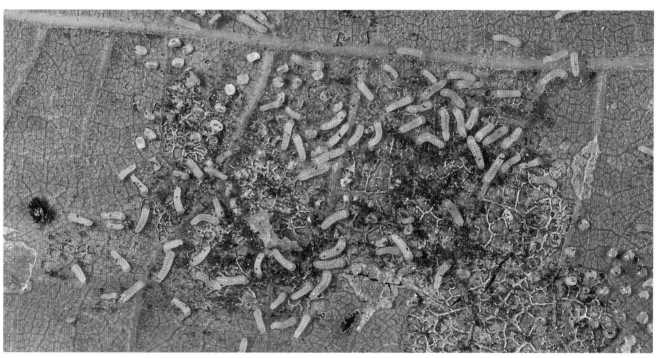

⊢
LENGTH

Western grapeleaf skeletonizer larvae normally feed side by side as first through early fourth instars, consuming entire areas of leaf. These first instars are infected with a granulosis virus that has disoriented them, causing larvae to wander, feed in scattered patches, and excrete dark diarrhea (Clausen 1961, 1978). Infected larvae discolor, grow abnormally, shrink, die, and drop from leaves. As

with Douglas fir tussock moth and some other moth species, when adult grapeleaf skeletonizers are infected, females scatter their eggs instead of laying them in batches. Many infected eggs fail to hatch. This virus is the most important microbial control agent of western grapeleaf skeletonizer. Larvae are also attacked by *Apanteles harrisinae* (Braconidae) and a parasitic fly (*Amedoria misella* or *Sturmia harrisinae*, Tachinidae). Photo by Max Badgley.

A citrus red mite infected by a virus specific to this citrus pest. Infected mites curl their legs under their body as shown. They may be able to walk stiffly but exhibit diarrhea and soon die. In dry weather, virus-infected mites dry up quickly; their shriveled bodies may be visible on plants or they may blow away, removing evidence of infection. Under humid conditions, infected mites disintegrate into reddish brown to black watery spots on fruit or leaves. If diseased mites are mounted on a slide and examined under a polarizing microscope, shiny internal crystals are evident (Smith and Cressman 1962).

1995), or predaceous beetles and bugs (Vasconcelos et al. 1996). Baculoviruses include nuclear polyhedrosis viruses (NPV) and granulosis viruses (GV) used to control pests. Nuclear polyhedrosis viruses get their name from the many rod-shaped virus particles that are embedded within each proteinaceous occlusion body when the virus is viewed through an electron microscope. In granulosis viruses, only one virus particle is present in each occlusion body.

As with bacteria, viruses usually must be eaten to infect their host. When insects are infected with virus, their bodies become soft, swollen, and dark or pale (table 9-2). Specific symptoms of virus infection vary depending on the host and the virus. Natural out-breaks of a virus can rapidly reduce pest populations under favorable conditions. For example, when a virus is present in a population of citrus red mites (usually under warm, moderately humid conditions) and mite densities exceed 3 or 4 adult females per leaf, a virus epidemic (epizootic) is likely to develop rapidly and drastically reduce mite populations. Similarly, naturally occurring outbreaks of NPV frequently decimate beet armyworm populations in cotton and other crops. Certain baculoviruses are commercially available (table 9-1), although successful use has so far been limited. Molecular biology techniques are being used to genetically engineer baculoviruses to increase their speed and efficacy (Hammock 1993, Hoover et al. 1995).

Resources

Many scientific journals, private organizations, government agencies, and universities regularly publish current information on biological pest control. Many of these resources can be identified by perusing the suggested reading and literature cited at the end of this book. Regularly printed resources on practical biological control information include:

California Agriculture
Division of Agriculture and
Natural Resources
University of California
1111 Franklin St.
Oakland, CA 94607-5200

The IPM Practitioner
Bio-Integral Resource Center (BIRC)
P.O. Box 7414
Berkeley, CA 94707
(510) 524-2567

Midwest Biological Control News
Department of Entomology
University of Wisconsin
1630 Linden Dr.
Madison, WI 53706

The World Wide Web (the Web) portion of the Internet is a vast source of information on biological control. The Web allows electronic access to university, government, industry, and private personal computers. Color photographs of pests and natural enemies, information on biology, management recommendations, decision-making models, and communication with biological control experts and practitioners are among the many resources available online through the Web. Although any printed list of online sites is immediately out of date, several relevant sites and their online address (or URL) at the time of this printing include:

IPM Project, University of
California Statewide
http://www.ipm.ucdavis.edu

Association of Natural
Bio-Control Producers
**http://ipmwww.ncsu.edu/biocontrol/
anbp/HomePage.html**

Biological Control Virtual
Information Center
**http://ipmwww.ncsu.edu/biocontrol/
biocontrol.html**

Cornell University Biological
Control Guide
http://nysaes.cornell.edu/ent/biocontrol

Suppliers of Beneficial Organisms
in North America
**http://www.cdpr.ca.gov/docs/dprdocs/
goodbug/organism.htm**

Sustainable Agriculture Research and Education Program, University of California
http://www.sarep.ucdavis.edu

USDA Biological Control Institute
http://www.aphis.usda.gov/oa/nbci/ nbci.html

SUPPLIERS

Many industry trade journals publish annual guides to suppliers of biological control organisms, monitoring equipment, selective pesticides, and other products used in biological and integrated pest control.

BioQuip Products
17803 LaSalle Ave.
Gardena, CA 90248-3602
(310) 324-0620

Gempler's
P.O. Box 270
Mt. Horeb, WI 53527
(800) 382-8473

Great Lakes IPM
10220 Church Road, NE
Vestaburg, MI 48891
(517) 268-5693

Suppliers of Beneficial Organisms in North America, is a free publication listing dozens of companies that sell natural enemies, available from:

California Department of Pesticide Regulation
1020 N Street, Rm 161
Sacramento, CA 95814-5604
(916) 324-4100
also online at: **http://www.cdpr.ca.gov/ docs/dprdocs/goodbug/organism.htm**

SUGGESTED READING

This suggested reading list includes both recent and outstanding earlier publications on biological control. See the literature cited for publications referenced in the text.

Anonymous. 1992. *Beyond Pesticides: Biological Approaches to Pest Management in California.* Oakland: Univ. Calif. Div. Agric. Nat. Res. Publ. 3354.

———. 1995. *Biologically Based Technologies For Pest Control.* Washington, D.C.: U. S. Congress, Office of Technology Assessment OTA-ENV-636.

———. 1996. *Ecologically Based Pest Management: New Solutions for a New Century.* Washington, D.C.: National Academy Press.

Askew, R. R. 1971. *Parasitic Insects.* New York: American Elsevier Publishing.

Borror, D. J., C. A. Triplehorn, and N. F. Johnson. 1989. *An Introduction to the Study of Insects.* 6th ed. Philadelphia: Saunders College.

Burgess, H. D., ed. 1981. *Microbial Control of Pests and Plant Diseases 1970-1980.* New York: Academic Press.

Clausen, C. P. 1940. *Entomophagous Insects.* New York: McGraw-Hill.

Clausen, C. P., ed. 1978. *Introduced Parasites and Predators of Arthropod Pests and Weeds: A World Review.* Washington, D.C.: U.S. Dept. Agric. Handb. 480.

Costello, R. A., D. P. Elliot, L. A. Gilkson, and D. R. Gillespie. 1992. *Integrated Control of Greenhouse Pests for Commercial Growers.* Victoria, BC: Ministry of Agriculture, Fisheries and Food.

Croft, B. A. 1990. *Arthropod Biological Control Agents and Pesticides.* New York: Wiley.

DeBach, P., ed. 1964. *Biological Control of Insect Pests and Weeds.* New York: Reinhold.

DeBach, P., and D. Rosen. 1991. *Biological Control by Natural Enemies.* 2d ed. New York: Cambridge University Press.

Dreistadt, S. H. 1994. *Pests of Landscape Trees and Shrubs: An Integrated Pest Management Guide.* Oakland: Univ. Calif. Div. Agric. Nat. Res. Publ. 3359.

Essig, E. O. 1926. *Insects of Western North America.* New York: Macmillan.

Flint, M. L. 1990. *Pests of the Garden and Small Farm: A Grower's Guide to Using Less Pesticide.* Oakland: Univ. Calif. Div. Agric. Nat. Res. Publ. 3332.

Flint, M. L., and R. van den Bosch. 1981. *Introduction to Integrated Pest Management.* New York: Plenum Press.

Harley, K. L. S., and I. W. Forno. 1992. *Biological Control of Weeds: A Handbook for Practitioners and Students.* Melbourne, Australia: Inkata Press.

Hoffmann, M. P., and A. C. Frodsham. 1993. *Natural Enemies of Vegetable Insect Pests.* Ithaca: Cornell University Cooperative Extension.

Hoffmann. M. P., C. H. Petzoldt, and A. C. Frodsham. 1996. *Integrated Pest Management for Onions.* Ithaca, NY: Cornell University Cooperative Extension.

Hoy, M. A., and D. C. Herzog, eds. 1985. *Biological Control in Agricultural IPM Systems.* New York: Academic Press.

Hoy, M. A., G. L. Cunningham, and L. Knutson, eds. 1987. *Biological Control of Pests by Mites.* Oakland: Univ. Calif. Div. Agric. Nat. Res. Publ. 3304.

Hussey, N. W., and N. Scopes, eds. 1985. *Biological Pest Control: The Glasshouse Experience.* Ithaca, NY: Cornell University Press.

Julien. M. H. 1992. *Biological Control of Weeds: A World Catalogue of Agents and Their Target Weeds.* 3d ed. Brisbane, Australia: CAB International.

Lumsden, R. D., and J. L. Vaughn, eds. 1993. *Pest Management: Biologically Based Technologies.* Conference Proceedings. Beltsville, MD: American Chemical Society.

MacKauer, M., L. E. Ehler, and J. Roland, eds. 1990. *Critical Issues in Biological Control*. Andover, UK: Intercept.

Mahr, D. L., and N. M. Ridgway. 1993. *Biological Control of Insects and Mites: An Introduction to Beneficial Natural Enemies and Their Use in Pest Management*. Madison: North Central Regional Coop. Ext. Publ. 481.

Mahr, S. E. R., D. L. Mahr, and J. Wyman. 1993. *Biological Control of Insect Pests of Cabbage and Other Crucifers*. Madison: North Central Regional Coop. Ext. Publ. 471.

Malais, M., and W. J. Ravensberg. 1992. *Knowing and Recognizing the Biology of Glasshouse Pests and Their Natural Enemies*. Berkel en Rodenrijs, Netherlands: Koppert Biological Systems.

Nechols, J. R., L. A. Andres, J. W. Beardsley, R. D. Goeden, and C. G. Jackson, eds. 1995. *Biological Control in the Western United States: Accomplishments and Benefits of Regional Project W-84, 1964–1989*. Oakland: Univ. Calif. Div. Agric. Nat. Res. Publ. 3361.

Olkowski, W., S. Daar, and H. Olkowski. 1991. *Common-Sense Pest Control*. Newtown, CT: Taunton Press.

Orloff, S. B., and H. L. Carlson. 1997. *Intermountain Alfalfa Management*. Oakland: Univ. Calif. Div. Agric. Nat. Res. Publ. 3366.

Raupp, M. J., R. G. Van Driesche, and J. A. Davidson. 1993. *Biological Control of Insect and Mite Pests of Woody Landscape Plants: Concepts, Agents and Methods*. College Park: University of Maryland.

Reuveni, R. 1995. *Novel Approaches to Integrated Pest Management*. Boca Raton, FL: Lewis Pub.

Steiner, M. Y., and D. P. Elliott. 1987. *Biological Pest Management for Interior Plantscapes*. 2d ed. Vegreville: Alberta Environmental Centre.

Sterling, G. R. 1991. *Biological Control of Plant Parasitic Nematodes: Progress, Problems, and Prospects*. Wallingford, UK: CAB International.

Tanada, Y., and H. K. Kaya. 1993. *Insect Pathology*. San Diego: Academic Press.

Thomson, W. T. 1992. *A Worldwide Guide to Beneficial Animals (Insects, Mites and Nematodes) Used for Pest Control Purposes*. Fresno: Thomson.

Tjamos, E. S., G. C. Papavizas, and R. J. Cook, eds. 1992. *Biological Control of Plant Diseases: Progress and Challenges for the Future*. New York: Plenum.

van den Bosch, R., P. S. Messenger, and A. P. Gutierrez. 1982. *An Introduction to Biological Control*. New York: Plenum.

Van Driesche, R. G., and T. S. Bellows. 1996. *Biological Control*. New York: Chapman and Hall.

Van Driesche, R., and E. Carey, eds. 1987. *Opportunities for Increased Use of Biological Control in Massachusetts*. Amherst: Univ. Mass. Agric. Exp. Sta. Res. Bull. 718.

Literature Cited

THESE PUBLICATIONS ARE CITED in the text. They are primarily recent secondary sources, review articles, and publications of practical use to pest managers. For primary literature and basic research, consult an expert, conduct a current literature search, and review references in the publications cited here.

Alomar, O., and R. N. Wiedenmann, eds. 1996. *Zoophytophagous Heteroptera: Implications for Life History and Integrated Pest Management*. Lanham, MD: Entomological Society of America.

Altieri, M. A. 1994. *Biodiversity and Pest Management in Agroecosystems*. Binghamton, NY: Haworth Press.

Andres, L. A., and E. M. Coombs. 1995. Scotch broom, *Cytisus scoparius* (L.) Link Fabaceae. In *Biological Control in the Western United States*. J. R. Nechols et al., eds. Oakland: Univ. Calif. Div. Agric. Nat. Res. Publ. 3361. 303–305.

Andres, L. A., and R. D. Goeden. 1995. Puncturevine, *Tribulus terrestris* L. Zygophyllaceae. In *Biological Control in the Western United States*. J. R. Nechols et al., eds. Oakland: Univ. Calif. Div. Agric. Nat. Res. Publ. 3361. 318–321.

Andres, L. A., and N. E. Rees. 1995. Musk thistle, *Carduus nutans* L. Asteraceae. In *Biological Control in the Western United States*. J. R. Nechols, et al., eds. Oakland: Univ. Calif. Div. Agric. Nat. Res. Publ. 3361. 248–251.

Andres, L. A., E. M. Coombs, and J. P. McCaffrey. 1995. Mediterranean sage, *Salvia aethiopis* L. Lamiaceae. In *Biological Control in the Western United States*. J. R. Nechols et al., eds. Oakland: Univ. Calif. Div. Agric. Nat. Res. Publ. 3361. 296–298.

Annecke, D. P. 1959. The effect of parathion and ants on *Coccus hesperidum* L. (Coccidae: Hemiptera) and its natural enemies. *J. Ent. Soc. So. Africa* 1:245–274.

Anonymous. 1952. *The Yearbook of Agriculture: Insects*. Washington, D.C.: U.S. Dept. Agric.

———. 1960. *The Elm Leaf Beetle*. Washington, D.C.: U.S. Dept. Agric. Leaflet 184.

———. 1979. *The Elm Leaf Beetle (Pyrrhalta luteola)*. Berkeley, CA: John Muir Institute.

———. 1985a. *Integrated Pest Management for Almonds*. Oakland: Univ. Calif. Div. Agric. Nat. Res. Publ. 3308.

———. 1985b. *Integrated Pest Management for Alfalfa Hay*. Oakland: Univ. Calif. Div. Agric. Nat. Res. Publ. 3312.

———. 1986. *Pesticide Resistance: Strategies and Tactics for Management*. Washington, D.C.: National Academy Press.

———. 1987. *Integrated Pest Management for Walnuts*. Oakland: Univ. Calif. Div. Agric. Nat. Res. Publ. 3270.

———. 1991a. *Integrated Pest Management for Apples and Pears*. Oakland: Univ. Calif. Div. Agric. Nat. Res. Publ. 3340.

———. 1991b. *Integrated Pest Management for Citrus*. 2d ed. Oakland: Univ. Calif. Div. Agric. Nat. Res. Publ. 3303.

———. 1995a. *Biologically Based Technologies for Pest Control*. Washington, D.C.: U.S. Congress, Office of Technology Assessment OTA-ENV-636.

———. 1995b. *BIOS for Almonds: A Practical Guide to Biological Integrated Orchard Systems Management*. Davis, CA: Community Alliance With Family Farmers Foundation.

———. 1996. *Integrated Pest Management for Cotton*. 2d ed. Oakland: Univ. Calif. Div. Agric. Nat. Res. Publ. 3305.

Arnaud, P. H. 1978. *A Host-Parasite Catalog of North American Tachinidae (Diptera)*. Washington, D.C.: U.S. Dept. Agric. Misc. Publ. 1319.

Asante, S. K. 1995. Functional response of European earwig and two species of coccinellids to densities of *Eriosoma lanigerum* (Hausmann) (Hemiptera: Aphididae). *J. Aust. Ent. Soc.* 34:105–109.

Ascerno, M. E. 1991. Insect phenology and integrated pest management. *J. Arbor.* 17:13–15.

Asteraki, E. J. 1993. The potential of carabid beetles to control slugs in grass/cover swards. *Entomophaga* 38:193–198.

Badii, M. H., and J. A. McMurtry. 1984. Life history of and life table parameters for *Phytoseiulus longipes* with comparative studies on *P. persimilis* and *Typhlodromus occidentalis* (Acari: Phytoseiidae). *Acarologia* 25:111–123.

Bai, B., S. Cobanoğlu, and S. M. Smith. 1995. Assessment of *Trichogramma* species for biological control of forest lepidopteran defoliators. *Entomologia Experimentalis et Applicata* 75:135–143.

Baker, T. C., S. E. Van Vorhis Key, and L. K. Gaston. 1985. Bait-preference tests for the Argentine ant (Hymenoptera: Formicidae). *J. Econ. Entomol.* 78:1083–1088.

Barbosa, P., and S. Braxton. 1993. A proposed definition of biological control and its relationship to related control approaches. In *Pest Management: Biologically Based Technologies*. R. D. Lumsden and J. L. Vaughn, eds. Conference Proceedings. Beltsville, MD: American Chemical Society. 21–27.

Barker, P. C., W. R. Bowen, V. E. Burton, C. S. Davis, A. S. Deal, L. D. McGrew, W. G. Schneeflock, W. R. Schreader, and J. E. Swift. 1978. *A Study of Insects: 4-H Entomology Project*. Oakland: Univ. Calif. Agric. Sci. Leaflet 2949.

Barr, B. A., G. W. Hickman, and C. S. Koehler. 1984. *Spiders*. Oakland: Univ. Calif. Div. Agric. Nat. Res. Leaflet 2531.

Bartlett, B. R. 1961. The influence of ants upon parasites, predators, and scale insects. *Ann. Entomol. Soc. Am.* 54:543–551.

Bartlett, B. R., and D. C. Lloyd. 1958. Mealybugs attacking citrus in California: Survey of their natural enemies and the release of new parasites and predators. *J. Econ. Entomol.* 51:90–93.

Baur, M. E., H. K. Kaya, and G. S. Thurston. 1995. Factors affecting entomopathogenic nematode infection of *Plutella xylostella* on a leaf surface. *Entomologia Experimentalis et Applicata* 77:239–250.

Bay, E. C., C. O. Berg, H. C. Chapman, and E. F. Legner. 1976. Biological control of medical and veterinary pests. In *Theory and Practice of Biological Control*. C. B. Huffaker and P. S. Messenger, eds. New York: Academic Press. 457–479.

Bell, C. E., J. N. Guerrero, and E. Y. Granados. 1996. A comparison of sheep grazing with herbicides for weed control in seedling alfalfa in the irrigated Sonoran Desert. *J. Production Agric.* 9:123–129.

Bellows, T. S., J. G. Morse, and L. K. Gaston. 1993. Residual toxicity of pesticides used for Lepidopteran insect control on citrus to *Aphytis melinus* DeBach (Hymenoptera: Aphelinidae). *Can. Ent.* 125:995–1001.

Bellows, T. S., J. G. Morse, D. G. Hadjidemetriou, and Y. Iwata. 1985. Residual toxicity of four insecticides used for control of citrus thrips (Thysanoptera: Thripidae) on three beneficial species in a citrus ecosystem. *J. Econ. Entomol.* 78:681–686.

Bennett, F. D., D. Rosen, P. Cochereau, and B. J. Wood. 1976. Biological control of pests of tropical fruits and nuts. In *Theory and Practice of Biological Control*. C. B. Huffaker and P. S. Messenger, eds. New York: Academic Press. 359–395.

Bentley, W. J., F. Zalom, J. Granett, and R. Smith. 1996. UC IPM Pest Management Guidelines: Grapes: Insects and Mites. UC IPM Guidelines Series 18. In *UC IPM Pest Management Guidelines*, Univ. Calif. Div. Agric. Nat. Res. Publ. 3339.

Bethell, R. S., R. B. Elkins, R. van Steenwyk, and L. Varela. 1995. Pear Pest Management Guidelines: Insects and Mites. UC IPM Guidelines Series 16. In *UC IPM Pest Management Guidelines*, Univ. Calif. Div. Agric. Nat. Res. Publ. 3339.

Bethke, J. A., and T. D. Paine. 1991. Screen hole size and barriers for exclusion of insect pests of glasshouse crops. *J. Entomol. Sci.* 26:169–177.

Bethke, J. A., R. A. Redak, and T. D. Paine. 1994. Screens deny specific pests entry to greenhouses. *Calif. Agric.* 48(3):37–40.

Blossey, B. 1995. Host specificity screening of insect biological control agents as part of an environmental risk assessment. In *Biological Control: Benefits and Risks*, H. M. T. Hokkanen and J. M. Lynch, eds. Cambridge: Cambridge University Press. 84–89.

Bond, A. A. 1976. *Hunger and Foraging in the Green Lacewing, Chrysopa carnea Stephens*. Ph.D. Diss. University of California, Berkeley.

Borror, D. J., D. M. De Long, and C. A. Triplehorn. 1981. *An Introduction to the Study of Insects*. 5th ed. Philadelphia: Saunders College.

Breene, R. G. 1995. *Common Names of Arachnids*. South Padre Island, TX: American Tarantula Society.

Breene, R. G., R. L. Meagher, D. A. Nordlund, and Y. Wang. 1992. Biological control of *Bemisia tabaci* (Homoptera: Aleyrodidae) in a greenhouse using *Chrysoperla rufilabris* (Neuroptera: Chrysopidae). *Biol. Control* 2:9–14.

Bridge, J. 1996. Nematode management in sustainable and subsistence agriculture. *Ann. Rev. Phytopath.* 34:201–225.

Bugg, R. L. 1993. Habitat manipulation to enhance the effectiveness of aphidophagous hover flies (Diptera: Syrphidae). *Sustainable Agric. Tech. Rev.* 5(2):12–15.

Bugg, R. L., and J. H. Anderson. n.d. *Farmscaping in California: Hedgerows, Native Grasses, Cover Crops, and Wild Plants for Biointensive Pest Management*. Davis: Univ. Calif. Sustainable Agriculture Education and Research Program

Bugg, R. L., and C. Waddington. 1994. Using cover crops to manage arthropod pests of orchards: A review. *Agric., Ecosys. and Env.* 50:11–28.

Bugg, R. L., and L. T. Wilson. 1989. *Ammi visnaga* (L.) Lamarck (Apiaceae): Associated beneficial insects and implications for biological control, with emphasis on the bell-pepper agroecosystem. *Biol. Agric. Hort.* 6:241–268.

Bugg, R. L., L. E. Ehler, and L. T. Wilson. 1987. Effect of common knotweed (*Polygonum aviculare*) on abundance and efficacy of insect predators of crop pests. *Hilgardia* 55:1–52.

Bugg, R. L., G. McGourty, M. Sarrantonio, W. T. Lanini, and R. Bartolucci. 1996. Comparison of 32 cover crops in an organic vineyard of the North Coast of California. *Biol. Agric. Hort.* 13:63–81

Bull, C. T., J. P. Stack, and J. L. Smilanick. 1997. *Pseudomonas syringae* strains ESC-10 and ESC-11 survive in wounds on citrus and control green and blue molds of citrus. *Biol. Control* 8:81–88.

Bull, B. C., M. J. Raupp, M. R. Hardin, and C. S. Sadof. 1993. Suitability of five horticulturally important armored scale insects as hosts for an exotic predaceous lady beetle. *J. Environ. Hort.* 11:28–30.

Burlando, T. M., H. K. Kaya, and P. Timper. 1993. Insect-parasitic nematodes are effective against black vine weevil. *Calif. Agric.* 47(3):16–18.

Buschman, L. L., W. H. Whitcomb, R. C. Hemenway, D. L. Mays, N. Ru, N. C. Leppla, and B. J. Smittle. 1977. Predators of velvetbean caterpillar eggs in Florida soybeans. *Environ. Entomol.* 6:403–407.

Calvet, C., J. Pera, and J. M. Barea. 1993. Growth response of marigold (*Tagetes erecta* L.) to inoculation with *Glomus mosseae*, *Trichoderma aureoviride* and *Pythium ultimum* in a peat-perlite mixture. *Plant Soil* 148:1–6.

Campbell, M. M. 1975. Duration of toxicity of residues of malathion and spray oil on citrus foliage in South Australia to adults of a California red scale parasite *Aphytis melinus* DeBach (Hymenoptera: Aphelinidae) *J. Aust. Ent. Soc.* 14:161–164.

Campbell, C. L., and J. P. McCaffrey. 1991. Population trends, seasonal phenology, and impact of *Chrysolina quadrigemina*, *C. hyperici* (Coleoptera: Chrysomelidae), and *Agrilus hyperici* (Coleoptera: Buprestidae) associated with *Hypericum perforatum* in northern Idaho. *Environ. Entomol.* 20:303–315.

Canard, M., Y. Séméria, and T. R. New. 1984. *Biology of Chrysopidae.* The Hague, Netherlands: Junk.

Caprile, J., C. Pickel, R. S. Bethell, and W. W. Barnett. 1996. Apple Pest Management Guidelines: Insects and Mites. UC IPM Guidelines Series 12. In *UC IPM Pest Management Guidelines*, Univ. Calif. Div. Agric. Nat. Res. Publ. 3339.

Caroli, L., I. Glazer, and R. Gaugler. 1996. Entomopathogenic nematode infectivity assay: Comparison of penetration rate into different hosts. *Biocontrol Sci. Tech.* 6:227–233.

Carroll, D. P., and S. C. Hoyt. 1984. Augmentation of European earwigs (Dermaptera: Forficulidae) for biological control of apple aphid (Homoptera: Aphididae) in an apple orchard. *J. Econ. Entomol.* 77:738–740.

Cartwright, D. K., and D. M. Benson. 1995. Biological control of Rhizoctonia stem rot of poinsettia in polyfoam rooting cubes with *Pseudomonas cepacia* and *Paecilomyces lilacinus*. *Biol. Control* 5:237–244.

Casale, W. L., V. Minassian, J. A. Menge, C. J. Lovatt, E. Pond, E. Johnson, and F. Guillemet. 1995. Urban and agricultural wastes for use as mulches on avocado and citrus and for delivery of microbial biocontrol agents. *J. Hort. Sci.* 70:315–332.

Castellano, M. A., and R. Molina. 1989. Mycorrhizae. In *The Container Tree Nursery Manual*. Vol. 5. T. D. Landis, R. W. Tinus, S. E. McDonald, and J. P. Barnett, eds. Washington, D.C.: U.S. Dept. Agric. Handb. 674.

Chilton, E. W., and M. I. Muoneke. 1992. Biology and management of grass carp (*Ctenopharyngodon idella*, Cyprinidae) for vegetation control: A North American perspective. *Reviews in Fish Biol. and Fisheries* 2:283–320.

Chittenden, F. H. 1912. *Insects Injurious to Vegetables*. New York: Orange Judd.

———. 1921. *The Beet Leaf-Beetle and its Control*. Washington, D.C.: U.S. Dept. Agric. Farmer's Bull. 1193.

Choo, H. Y., A. M. Koppenhöfer, and H. K. Kaya. 1996. Combination of two entomopathogenic nematode species for suppression of an insect pest. *J. Econ. Entomol.* 89:97–103.

Clausen, C. P. 1961. Biological control of western grape leaf skeletonizer (*Harrisina brilliams* B. and McD.) in California. *Hilgardia* 31:613–637.

———. 1978. *Introduced Parasites and Predators of Arthropod Pests and Weeds: A World Review*. Washington, D.C.: U.S. Dept. Agric. Handb. 480.

Coderre, D., E. Lucas, and I. Gagné. 1995. The occurrence of *Harmonia axyridis* (Pallas) (Coleoptera: Coccinellidae) in Canada. *Can. Ent.* 127:609–611.

Cole, F. R. 1969. *The Flies of Western North America*. Berkeley: University of California Press.

Coll, M., and R. Ridgway. 1995. Functional and numerical responses of *Orius insidiosus* (Heteroptera: Anthocoridae) to its prey in different vegetable crops. *Ann. Entomol.* 88:732–738.

Compere, H., and H. S. Smith. 1932. The control of the citrophilus mealybug, *Pseudococcus gahani*, by Australian parasites. *Hilgardia* 6:585–618.

Cook, R. J, W. L. Bruckart, J. R. Coulson, M. S. Goettel, R. A. Humber, R. D. Lumsden, J. V. Maddox, M. L. McManus, L. Moore, S. F. Meyer, P. C. Quimby, J. P. Stark, and J. L. Vaughn. 1996. Safety of microorganisms intended for pest and plant disease control: A framework for scientific evaluation. *Biol. Control* 7:333–351.

Corbett, A., and R. E. Plant. 1993. Role of movement in the response of natural enemies to agroecosystem diversification: A theoretical evaluation. *Environ. Entomol.* 22:519–531.

Corbett, A., and J. A. Rosenheim. 1996. Impact of a natural enemy overwintering refuge and its interaction with the surrounding landscape. *Ecol. Entomol.* 21:155–164.

Costello, M. J., M. A. Mayse, K. M. Daane, W. A. O'Keefe, and C. B. Sisk. 1995. *Spiders in San Joaquin Valley Grape Vineyards*. Oakland: Univ. Calif. Div. Agric. Nat. Res. Leaflet 21530.

Coulson, J. R., R. S. Soper, and D. W. Williams, eds. 1991. *Biological Control Quarantine Needs and Procedures*. Washington, D.C.: U.S. Dept. Agric. Res. Serv. Proceedings ARS-99.

Coviello, R., W. J. Bentley, and W. W. Barnett. 1996. Fig Pest Management Guidelines: Insects and Mites. UC IPM Guidelines Series 4. In *UC IPM Pest Management Guidelines*, Univ. Calif. Div. Agric. Nat. Res. Publ. 3339.

Cowles, R., and K. Kido. 1996. UC IPM Pest Management Guidelines: Turfgrass: Insects and Mites. UC IPM Landscape and Ornamental Series 1. In *UC IPM Pest Management Guidelines, Turfgrass*, Univ. Calif. Div. Agric. Nat. Res. Publ. 3365-T.

Cowles, R. S., and J. A. Downer. 1995. Eucalyptus snout beetle detected in California. *Calif. Agric.* 49(1):38–40.

Cranshaw, W., D. C. Sclar, and D. Cooper. 1996. A review of 1994 pricing and marketing by suppliers of organisms for biological control of arthropods in the United States. *Biol. Control* 6:291–296.

Crawford, H. S., and D. T. Jennings. 1989. Predation by birds on spruce budworm *Choristoneura fumiferana*: Functional, numerical, and total responses. *Ecology* 70:152–163.

Creamer, N. G., M. A. Bennett, B. R. Stinner, J. Cardina, and E. E. Regnier. 1996. Mechanisms of weed suppression in cover crop-based production systems. *HortScience* 31:410–413.

Croft, B. A. 1990. *Arthropod Biological Control Agents and Pesticides*. New York: Wiley.

Croft, B. A., and I. V. MacRae. 1992. Biological control of apple mites by mixed populations of *Metaseiulus occidentalis* and *Typhlodromus pyri* (Acari: Phytoseiidae). *Environ. Entomol.* 21:202–209.

Cushing, N., and M. E. Whalon. 1986. *Rearing Predator Mites for Orchards and Glasshouses*. Ann Arbor: Mich. State Univ. Coop. Ext. Serv. Publ. E-1872.

Daane, K. M., and L. E. Caltagirone. 1989. Biological control of black scale in olives. *Calif. Agric.* 43(1):9–11.

Daane, K. M., M. S. Barzman, C. E. Kennett, and L. E. Caltagirone. 1991. Parasitoids of black scale in California: Establishment of *Prococcophagus probus* Annecke & Mynhardt and *Coccophagus rusti* Compere (Hymenoptera: Aphelinidae) in olive orchards. *Pan-Pac. Entomol.* 67:99–106.

Daane, K. M., G. Y. Yokota, Y. Rasmussen, Y. Zheng, and K. S. Hagen. 1993. Effectiveness of leafhopper control varies with lacewing release methods. *Calif. Agric.* 47(6):19–23.

Daane, K. M., G. Y. Yokota, Y. Zheng, and K. S. Hagen. 1996. Inundative release of common green lacewings (Neuroptera: Chrysopidae) to suppress *Erythroneura variabilis* and *E. elegantula* (Homoptera: Cicadellidae) in vineyards. *Environ. Entomol.* 25: 1224–1234.

Dahlsten, D. L., W. A. Copper, D. L. Rowney, and P. K. Kleintjes. 1990. Quantifying bird predation of arthropods in forests. *Studies in Avian Biology* 13:44–52.

Dahlsten, D. L., S. M. Tait, D. L. Rowney, and B. J. Gingg. 1993. A monitoring system and development of ecologically sound treatments for elm leaf beetle. *J. Arbor.* 19:181–186.

Dahlsten, D. L., D. M. Kent, D. L. Rowney, W. A. Copper, T. E. Young, and R. L. Tassan. 1995. Parasitoid shows potential for biocontrol of eugenia psyllid. *Calif. Agric.* 49(4):36–40.

Dahlsten, D. L., D. L. Rowney, R. L. Tassan, W. A. Copper, W. E. Chaney, R. L. Robb, S. Tjosvold, M. Bianchi, and P. Lane. 1996. *Blue Gum Psyllid*. Novato: Univ. Calif. Coop. Exten. HortScript 20.

Daly, H. V., J. T. Doyen, and P. R. Ehrlich. 1978. *Introduction to Insect Biology and Diversity*. New York: McGraw-Hill.

Datnoff, L. E., S. Nemec, and K. Pernezny. 1995. Biological control of Fusarium crown and root rot of tomato in Florida using *Trichoderma harzianum* and *Glomus intraradices*. *Biol. Control* 5:427–431.

Davidson, N. A., J. E. Dibble, M. L. Flint, P. J. Marer, and A. Guye. 1991. *Managing Insects and Mites with Spray Oils*. Oakland: Univ. Calif. Div. Agric. Nat. Res. Publ. 3347.

Davies, V. T. 1986. *Australian Spiders Araneae: Collection, Preservation and Identification*. South Brisbane: Queensland Museum Booklet 14.

Davis, R. M., J. J. Nunez, R. N. Vargas, B. L. Weir, S. D. Wright, and D. J. Munier. 1996. Metam-sodium kills beneficial soil fungi as well as cotton pests. *Calif. Agric.* 50(5):42–44.

DeBach, P., ed. 1964. *Biological Control of Insect Pests and Weeds*. New York: Reinhold.

DeBach, P., and D. Rosen. 1991. *Biological Control by Natural Enemies*. 2d ed. New York: Cambridge University Press.

Denmark, H. A., and E. Schicha. 1983. Revision of the genus *Phytoseiulus* Evans (Acarina: Phytoseiidae). *Internat. J. Acarol.* 9:27–35.

Donkin, R. A. 1977. Spanish Red: An Ethnogeographical Study of Cochineal and the Opuntia Cactus. *Trans. Am. Philosophical Soc.* 67(5):1–84.

Dreistadt, S. H. 1994. *Pests of Landscape Trees and Shrubs: An Integrated Pest Management Guide*. Oakland: Univ. Calif. Div. Agric. Nat. Res. Publ. 3359.

Dreistadt, S. H., and D. L. Dahlsten. 1990a. Insecticide bark bands and control of the elm leaf beetle (Coleoptera: Chrysomelidae) in northern California. *J. Econ. Entomol.* 83:1495–1498.

———. 1990b. Distribution and abundance of *Erynniopsis antennata* (Dip.: Tachinidae) and *Tetrastichus brevistigma* (Hym.: Eulophidae), two introduced elm leaf beetle parasitoids in northern California. *Entomophaga* 35(4):527–536.

Dreistadt, S. H., and M. L. Flint. 1995. Ash whitefly (Homoptera: Aleyrodidae) overwintering and biological control by *Encarsia inaron* (Hymenoptera: Aphelinidae) in northern California. *Environ. Entomol.* 24:459–464.

———. 1996. Melon aphid (Homoptera: Aphididae) control by inundative convergent lady beetle (Coleoptera: Coccinellidae) release on chrysanthemum. *Environ. Entomol.* 25:688–697.

Dreistadt, S. H., and K. S. Hagen. 1994. Classical biological control of acacia psyllid, *Acizzia uncatoides* (Homoptera: Psyllidae), and predator-prey-plant interactions in the San Francisco Bay Area. *Biol. Control* 4:319–327.

Dreistadt, S. H., M. P. Parrella, and M. L. Flint. 1992. Integrating chemical and biological control of ornamental pests. In *Proceedings of the 1992 First Annual Southwest Ornamental IPM Workshop*, C. R. Ward, ed. Albuquerque: New Mexico State University. 26–41.

Duelli, P. 1980. Adaptive dispersal and appetitive flight in the green lacewing, *Chrysoperla carnea*. *Ecol. Entomol.* 5:213–220.

Easterbrook, M. A. 1992. The possibilities for control of two-spotted spider mite *Tetranychus urticae* on field-grown strawberries in the UK by predatory mites. *Biocontrol Sci. Tech.* 2:235–245.

Ehler, L. E. 1977. Natural enemies of cabbage looper on cotton in the San Joaquin Valley. *Hilgardia* 45:73–106.

———. 1995. Biological control of obscure scale (Homoptera: Diaspididae) in California: An experimental approach. *Environ. Entomol.* 24:779–795.

Ehler, L. E., and M. G. Kinsey. 1995. Ecology and management of *Mindarus kinseyi* Voegtlin (Aphidoidea: Mindaridae) on white-fir seedlings at a California forest nursery. *Hilgardia* 62(1):1–62.

Ehler, L. E., and R. van den Bosch. 1974. An analysis of the natural biological control of *Trichoplusia ni* (Lepidoptera: Noctuidae) on cotton in California. *Can. Ent.* 106:1067–1073.

Elmore, C. L., J. J. Stapleton, C. E. Bell, and J. E. DeVay. 1997. *Soil Solarization: A Nonpesticidal Method for Controlling Diseases, Nematodes, and Weeds*. Oakland: Univ. Calif. Div. Agric. Nat. Res. Leaflet 21377.

Emerton, J. H., and S. W. Frost. 1961. *The Common Spiders of the United States*. New York: Dover.

Entwistle, P. F., A. C. Forkner, B. M. Green, and J. S. Cory. 1993. Avian dispersal of nuclear polyhedrosis viruses after induced epizootics in the pine beauty moth, *Panolis flammea* (Lepidoptera: Noctuidae). *Biol. Control* 3:61–69.

Essig, E. O. 1926. *Insects of Western North America*. New York: Macmillan.

Evans, E. W., and J. G. Swallow. 1993. Numerical responses of natural enemies to artificial honeydew in Utah alfalfa. *Environ. Entomol.* 22:1392–1401.

Fan, Y., and F. L. Petitt. 1994. Biological control of broad mite, *Polyphagotarsonemus latus* (Banks), by *Neoseiulus barkeri* Hughes on pepper. *Biol. Control* 4:390–395.

Fang, J. G., and P. H. Tsao. 1995. Efficacy of *Penicillium funiculosum* as a biological control agent against Phytophthora root rots of azalea and citrus. *Phytopath.* 85:871–878.

Farr, D. F., G. F. Bills, G. P. Chamuris, and A. Y. Rossman. 1989. *Fungi on Plants and Plant Products in the United States*. St. Paul, MN: American Phytopathological Society.

Feener, D. H., and B. V. Brown. 1997. Diptera as parasitoids. *Ann. Rev. Entomol.* 42: 73–97.

Ferris, H., W. V. Masuda, C. E. Castro, E. P. Caswell, B. A. Jaffee, P. A. Roberts, B. B. Westerdahl, and V. M. Williamson. 1992. Biological approaches to management of plant-parasitic nematodes. In *Beyond Pesticides: Biological Approaches to Pest Management in California*. Oakland: Univ. Calif. Div. Agric. Nat. Res. Publ. 3354. 68–101.

Field, R. P., and M. A. Hoy. 1984. Biological control of spider mites on greenhouse roses. *Calif. Agric.* 38(3/4):29–32.

Fisher, T. W. 1963. *Mass culture of Cryptolaemus and Leptomastix—Natural Enemies of Citrus Mealybug*. Univ. Calif. Agric. Exp. Sta. Bull. 797.

Flaherty, D. L., and C. B. Huffaker. 1970. Biological control of Pacific mites and Willamette mites in San Joaquin Valley orchards. *Hilgardia* 40:267–330.

Flanders, S. E. 1940. Biological control of the long-tailed mealybug, *Pseudococcus longispinus*. *J. Econ. Entomol.* 33:754–759.

Flinn, P. W., D. W. Hagstrum, and W. H. McGaughey. 1996. Suppression of beetles in stored wheat by augmentative release of parasitic wasps. *Environ. Entomol.* 25:505–511.

Flint, M. L. 1990. *Pests of the Garden and Small Farm: A Grower's Guide to Using Less Pesticide*. Oakland: Univ. Calif. Div. Agric. Nat. Res. Publ. 3332.

Flint, M. L., and P. A. Roberts. 1988. Using crop diversity to manage pest problems: Some California examples. *Am. J. of Alt. Agric.* 3:163–167.

Flint, M. L., and R. van den Bosch. 1981. *Introduction to Integrated Pest Management*. New York: Plenum Press.

Flint, M. L., T. C. Baker, D. L. Dahlsten, B. A. Federici, T. W. Fisher, G. Gordh, J. Henry, H. K. Kaya, J. G. Miller, G. O. Poinar, and P. V. Vail. 1992. Biological approaches to management of arthropods. In *Beyond Pesticides: Biological Approaches to Pest Management in California*. Oakland: Univ. Calif. Div. Agric. Nat. Res. Publ. 3354. 2–67.

Float, K. D., and T. G. Whitham. 1994. Ant-aphid interaction reduces chrysomelid herbivory in a cottonwood hybrid zone. *Oecologia* 97:215–221.

Forster, L. D., R. F. Luck, and E. E. Grafton-Cardwell. 1995. *Life Stages of California Red Scale and Its Parasitoids*. Oakland: Univ. Calif. Div. Agric. Nat. Res. Publ. 21529.

Frank, W. 1943. The entomological control of St. Johns wort (*Hypericum perforatum L.*) with particular reference to the insect enemies of the weed in southern France. *Austral. Council Sci. Indus. Res. Bull.* 169:1–88.

Fukui, R., M. N. Schroth, M. Hendson, and J. G. Hancock. 1994a. Interaction between strains of pseudomonads in sugar beet spermospheres and their relationship to pericarp colonization by *Pythium ultimum* in soil. *Phytopath.* 84:1322–1330.

Fukui, R., M. N. Schroth, M. Hendson, J. G. Hancock, and M. K. Firestone. 1994b. Growth patterns and metabolic activity of Pseudomonads in sugar beet spermospheres: relationship to pericarp colonization by *Pythium ultimum*. *Phytopath.* 84:1331–1338.

Gagné, R. J. 1989. *The Plant-Feeding Gall Midges of North America*. Ithaca, NY: Cornell University Press.

Gamliel, A., and J. J. Stapleton. 1993. Effect of soil amendment with chicken compost or ammonium phosphate and solarization on pathogen control, rhizosphere microorganisms, and lettuce growth. *Plant Dis.* 77:886–891.

Garcia, R., L. E. Caltagirone, and L. E. Gutierrez. 1988. Comments on a redefinition of biological control. *BioSci.* 3:692–694.

Gardner, J., and K. Giles. 1996a. Handling and environmental effects on viability of mechanically dispensed green lacewing eggs. *Biol. Control* 7:245–250.

———. 1996b. Mechanical distribution of *Chrysoperla rufilabris* and *Trichogramma pretiosum*: Survival and uniformity of discharge after spray dispersal in an aqueous suspension. *Biol. Control* 8:138–142.

Gaugler, R., and H. K. Kaya. 1990. *Entomopathogenic Nematodes in Biological Control*. Boca Raton, FL: CRC Press.

Gelernter, W. D. 1996. Insect degree-day models for turf. *Golf Course Mgmt.* 64:63–72.

Gill, S., J. A. Davidson, and M. J. Raupp. 1992. Control of peachtree borer using entomopathogenic nematodes. *J. Arbor.* 18:184–187.

Gill, S., J. Davidson, W. MacLachlan, and W. Potts. 1994. Controlling banded ash clearwing moth borer using entomopathogenic nematodes. *J. Arbor.* 20:146–149.

Giroux, S., R. M. Duchesne, and D. Coderre. 1995. Predation of *Leptinotarsa decemlineata* (Coleoptera: Chrysomelidae) by *Coleomegilla maculata* (Coleoptera: Coccinellidae): Comparative effectiveness of predator developmental stages and effect of temperature. *Environ. Entomol.* 24:748–754.

Glofcheskie, B. D., and G. A. Gordon. 1993. Efficacy of Muscovy ducks as an adjunct for house fly (Diptera: Muscidae) control in swine and dairy operations. *J. Econ. Entomol.* 86:1686–1692.

Godan, D. 1983. *Pest Slugs and Snails: Biology and Control*. Berlin: Springer-Verlag.

Godfrey, L. D. 1995. Dry Bean Pest Management Guidelines: Insects. UC IPM Guidelines Series 19. In *UC IPM Pest Management Guidelines*, Univ. Calif. Div. Agric. Nat. Res. Publ. 3339.

Godfrey, L. D., and T. F. Leigh. 1994. Alfalfa harvest strategy effect on lygus bug (Hemiptera: Miridae) and insect predator populations density: Implications for use as trap crop in cotton. *Environ. Entomol.* 23:1106–1118.

Godfrey, L. D., R. L. Coviello, W. J. Bentley, C. G. Summers, J. J. Stapleton, M. Murray, and E. T. Natwick. 1996. *UC IPM Pest Management Guidelines: Cucurbits: Insects and Mites*. UC IPM Guidelines Series 27. In *UC IPM Pest Management Guidelines*, Univ. Calif. Div. Agric. Nat. Res. Publ. 3339.

Goeden, R. D. 1992. University perspectives on regulatory issues involving parasitic, predaceous, and phytophagous arthropods as biological control agents. In *Regulations and Guidelines: Critical Issues in Biological Control*, R. Charudattan and H. W. Browning, eds. Proceedings of a USDA/CSRS National Workshop, Vienna, Virginia, l0-12 June 1992. Institute of Food and Agricultural Sciences. Gainsville: University of Florida. 107–114.

———. 1995a. Italian thistle, *Carduus pycnocephalus* L. Asteraceae. In *Biological Control in the Western United States*. J. R. Nechols et al., eds. Oakland: Univ. Calif. Div. Agric. Nat. Res. Publ. 3361. 242–244.

———. 1995b. Milk thistle, *Silybum marianum* (L.) Gaertner Asteraceae. In *Biological Control in the Western United States*. J. R. Nechols et al., eds. Oakland: Univ. Calif. Div. Agric. Nat. Res. Publ. 3361. 245–247.

Goeden, R. D., and R. W. Pemberton. 1995. Russian thistle, *Salsola australis* R. Brown Chenopodiaceae. In *Biological Control in the Western United States*. J. R. Nechols et al., eds. Oakland: Univ. Calif. Div. Agric. Nat. Res. Publ. 3361. 276–280.

Goeden, R. D., C. A. Fleschner, and D. W. Ricker. 1967. Biological control of prickly pear cacti on Santa Cruz Island, California. *Hilgardia* 38:579–606.

Gordh, G., J. B. Woolley, and R. A. Medved. 1983. Biological studies on *Goniozus legneri* Gordh (Hymenoptera: Bethylidae), a primary external parasite of the navel orangeworm *Amyelois transitella* and pink bollworm *Pectinophora gossypiella* (Lepidoptera: Pyralidae, Gelechiidae). *Contrib. Am. Entomol.* 20:433–470.

Gordon, R. D. 1985. The Coccinellidae (Coleoptera) of America north of Mexico. *J. New York Entomol. Soc.* 93:1–912.

Gordon, R. D., and N. Vandenberg. 1991. Field guide to recently introduced species of Coccinellidae (Coleoptera) in North America, with a revised key to North American genera of Coccinellini. *Proc. Entomol. Soc. Wash.* 93:845–864.

———. 1995. Larval systematics of North American *Coccinella* L. (Coleoptera: Coccinellidae). *Ent. Scand.* 26:67–86.

Gorham, J. R., ed. 1991. *Insect and Mite Pests in Food: An Illustrated Key*. Vol. 2. Washington, D.C.: U.S. Dept. Agric. Handb. 655.

Gould, F., G. Kennedy, and R. Kopanic. 1996. Environmental issues associated with enhancing the impact of biological control agents: a student debate. *Am. Entomologist* 42:160–173.

Gould, J. R., T. S. Bellows, and T. D. Paine. 1992. Evaluation of biological control of *Siphoninus phillyreae* (Haliday) by the parasitoid *Encarsia partenopea* (Walker), using life-table analysis. *Biol. Control* 2:257–265.

Goulet, H., and J. T. Huber. 1993. *Hymenoptera of the World: An Identification Guide*. Ottawa: Agriculture Canada Research Branch Publ. 1894/E.

Graebner, L., D. S. Moreno, and J. L. Baritelle. 1984. The Fillmore Citrus Protective District: A success story in integrated pest management. *Bull. Ent. Soc. Amer.* 30:27–33.

Grafton-Cardwell, E. E., and Y. Ouyang. 1995a. Augmentation of *Euseius tularensis* (Acari: Phytoseiidae) in Citrus. *Environ. Entomol.* 24:738–747.

———. 1995b. Manipulation of the predaceous mite, *Euseius tularensis* (Acari: Phytoseiidae), with pruning for citrus thrips control. In *Thrips Biology and Management*, B. L. Parker, M. Skinner, and T. Lewis, eds. New York: Plenum. 251–254.

Grafton-Cardwell, E. E., A. Eller, and N. V. O'Connell. 1995. Integrated citrus thrips control reduces secondary pests. *Calif. Agric.* 49(2):23–28.

Grafton-Cardwell, E. E., J. G. Morse, N. V. O'Connell, and P. A. Phillips. 1996. UC IPM Pest Management Guidelines: Citrus: Insects, Mites, and Snails. UC IPM Guidelines Series 28. In *UC IPM Pest Management Guidelines*, Univ. Calif. Div. Agric. Nat. Res. Publ. 3339.

Grasswitz, T. R., and E. C. Burts. 1995. Effect of native natural enemies and augmentative release of *Chrysoperla rufilabris* Burmeister and *Aphidoletes aphidimyza* (Rondani) on the population dynamics of the green apple aphid, *Aphis pomi* De Geer. *Internat. J. Pest Mgmt.* 41:176–183.

Grasswitz, T. R., and T. D. Paine. 1992. Kairomonal effect of an aphid cornicle secretion on *Lysiphlebus testaceipes* (Cresson) (Hymenoptera: Aphidiidae). *J. Ins. Beh.* 5:447–457.

———. 1993. Influence of physiological state and experience on the responsiveness of *Lysiphlebus testaceipes* (Cresson) (Hymenoptera: Aphidiidae) to aphid honeydew and host plants. *J. Ins. Beh.* 6:511–528.

Greathead, D. J. 1995. Benefits and risks of classical biological control. In *Biological Control: Benefits and Risks*, H. M. T. Hokkanen and J. M. Lynch, eds. Cambridge: Cambridge University Press. 53–63.

Greathead, D. J., and A. H. Greathead. 1992. Biological control of insect pests by parasitoids and predators: BIOCAT database. *Biocontrol News Info.* 13:61N–68N.

Grissell. E. R., and M. E. Schauff. 1990. *A Handbook of the Families of Nearctic Chalcidoidea*. Washington, D.C.: Entomological Society of Washington.

Hagen, K. S. 1962. Biology and ecology of predaceous Coccinellidae. *Ann. Rev. Entomol.* 7:289–326.

———. 1974. The significance of predaceous Coccinellidae in biological and integrated control of insects. *Entomophaga* 7:25–44.

———. 1982. *Notes on the Convergent Lady Beetle (Hippodamia convergens).* Oakland: Univ. Calif. Agric. Sci. Leaflet 2502.

———. 1986. Ecosystem analysis: Plant cultivars (HPR), entomophagous species and food supplements. In *Interactions of Plant Resistance and Parasitoids and Predators*, D. J. Boethel and R. D. Eikenbary, eds. Chichester, UK: Ellis Horwood. 157–197.

Hagen, K. S., E. F. Sawall, and R. L. Tassan. 1971. The use of food sprays to increase effectiveness of entomophagous insects. *Proc. Tall Timbers Conference on Ecological Animal Control by Habitat Management* 3:59–81.

Hagler, J. R., and S. E. Naranjo. 1994. Qualitative survey of two coleopteran predators of *Bemisia tabaci* (Homoptera: Aleyrodidae) and *Pectinophora gossypiella* (Lepidoptera: Gelechiidae) using a multiple prey gut content ELISA. *Environ. Entomol.* 23:193–197.

Hajek, A. E. 1994. Field identification of the gypsy moth nuclear polyhedrosis virus (NPV). Durham, NH: U.S. Dept. Agric. Forest Service NA-PR-01-94.

Hajek, A. E., and D. W. Roberts. 1992. Field diagnosis of gypsy moth (Lepidoptera: Lymantriidae) larval mortality caused by *Entomophaga maimaiga* and the gypsy moth nuclear polyhedrosis virus. *Environ. Entomol.* 21:706–713.

Hajek, A. E., and A. L. Snyder. 1992. Field identification of the gypsy moth fungus, *Entomophaga maimaiga*. Durham, NH: U.S. Dept. Agric. Forest Service NA-PR-02-92.

Hajek, A. E., and R. J. St. Leger. 1994. Interactions between fungal pathogens and insect hosts. *Ann. Rev. Entomol.* 39:293–322.

Hajek, A. E., R. A. Humber, and J. S. Elkinton. 1995. Mysterious origin of *Entomophaga maimaiga* in North America. *American Entomologist* 41:31–42.

Hall, R. W., and L. E. Ehler. 1979. Rate of establishment of natural enemies in classical biological control. *Bull. Entomol. Soc. Amer.* 25:280–282.

Hammock, B. D. 1993. Recombinant baculoviruses as biological insecticides. In *Pest Management: Biologically Based Technologies*, R. D. Lumsden and J. L. Vaughn, eds. Conference Proceedings. Beltsville, MD: American Chemical Society. 313–325.

Haney, P. B., R. F. Luck, and D. S. Moreno. 1987. Increases in densities of citrus red mite, *Panonychus citri* (Acarina: Tetranychidae), in association with Argentine ant, *Iridomyrmex humilis* (Hymenoptera: Formicidae), in Southern California citrus. *Entomophaga* 32:49–57.

Haney, P., P. A. Phillips, and R. Wagner. 1987. *A Key to the Most Common and/or Economically Important Ants of California with Color Pictures.* Oakland: Univ. Calif. Div. Agric. Nat. Res. Leaflet 21433.

Haney, P. B., J. G. Morse, R. F. Luck, H. J. Griffiths, E. E. Grafton-Cardwell, and N. V. O'Connell. 1992. *Reducing Insecticide Use and Energy Costs in Citrus Pest Management.* Univ. Calif. Div. Agric. Nat. Res. UC IPM Publ. 15.

———. 1993. The rationale for adopting IPM. *Citrograph* 78(4):7, 9–12.

Hanks, L. M., T. D. Paine, and J. G. Miller. 1996. Tiny wasp helps protect eucalypts from eucalyptus longhorned borer. *Calif. Agric.* 50(3):14–16.

Hanks, L. M., J. R. Gould, T. D. Paine, J. G. Miller, and Q. Wang. 1995. Biology and host relations of *Avetianella longoi* (Hymenoptera: Encyrtidae), an egg parasitoid of the eucalyptus longhorned borer (Coleoptera: Cerambycidae). *Ann. Entomol.* 88:666–671.

Hara, A. H., H. K. Kaya, R. Gaugler, L. M. Lebeck, and C. L. Mello. 1993. Entomopathogenic nematodes for biological control of the leafminer, *Liriomyza trifolii* (Dipt.: Agromyzidae). *Entomophaga* 38:359–369.

Hare, J. D., and R. F. Luck. 1994. Environmental variation in physical and chemical cues used by the parasitic wasp, *Aphytis melinus*, for host recognition. *Entomologia Experimentalis et Applicata* 72:97–108.

Harley, K. L. S., and I. W. Forno. 1992. *Biological Control of Weeds: A Handbook for Practitioners and Students.* Melbourne, Australia: Inkata Press.

Harris, M. A., R. D. Oetting, and W. A. Gardner. 1995. Use of entomopathogenic nematodes and a new monitoring technique for control of fungus gnats, *Bradysia coprophila* (Diptera: Sciaridae), in floriculture. *Biol. Control* 5:412–418.

Hassan, S. A., F. Bigler, H. Bogenschütz, E. Boller, J. Brun, J. N. M. Calis, J. Coremans-Pelseneer, C. Duso, A. Grove, U. Heimbach, N. Helyer, H. Hokkanen, G. B. Lewis, F. Mansour, L. Moreth, L. Polgar, L. Samsøe-Petersen, B. Sauphanor, A. Stäubli, G. Sterk, A. Vainio, M. van de Veire, G. Viggiani, and H. Vogt. 1994. Results of the sixth joint pesticide testing programme of the IOBC/WPRS working group–pesticides and beneficial organisms. *Entomophaga* 39:107–119.

Hattingh, V., and M. J. Samways. 1994. Physiological and behavioral characteristics of *Chilocorus* spp. (Coleoptera: Coccinellidae) in the laboratory relative to effectiveness in the field as biocontrol agents. *J. Econ. Entomol.* 87:31–38.

Heinz, K. M., and J. M. Nelson. 1996. Interspecific interactions among natural enemies of *Bemisia* in an inundative biological control program. *Biol. Control* 6:384–393.

Heinz, K. M., and M. P. Parrella. 1990. Biological control of insect pests on greenhouse marigolds. *Environ. Entomol.* 19:825–835.

———. 1994. Biological control of *Bemisia argentifolii* (Homoptera: Aleyrodidae) infesting *Euphorbia pulcherrima*: Evaluations of releases of *Encarsia luteola* (Hymenoptera: Aphelinidae) and *Delphastus pusillus* (Coleoptera: Coccinellidae). *Environ. Entomol.* 23:1346–1353.

Heinz, K. M., J. P. Newman, and M. P. Parrella. 1988. Biological control of leafminers on greenhouse marigolds. *Calif. Agric.* 42(2):10–12.

Heinz, K. M., L. Nunney, and M. P. Parrella. 1993. Toward predictable biological control of *Liriomyza trifolii* (Diptera: Agromyzidae) infesting greenhouse chrysanthemums. *Environ. Entomol.* 22:1217–1233.

Heinz, K. M., B. F. McCutchen, R. Herrmann, M. P. Parrella, and B. D. Hammock. 1995. Direct effects of recombinant nuclear polyhedrosis viruses on selected nontarget organisms. *J. Econ. Entomol.* 88:259–264.

Hendricks, L. C., W. W. Barnett, M. M. Barnes, C. Pickel, W. H. Olson, G. S. Sibbett, and R. A. van Steenwyk. 1996. Walnut Pest Management Guidelines: Insects and Mites. UC IPM Guidelines Series 2. In *UC IPM Pest Management Guidelines*, Univ. Calif. Div. Agric. Nat. Res. Publ. 3339.

Henry, T. J., and R. C. Froeschner. 1988. *Catalog of the Heteroptera, or True Bugs, of Canada and the Continental United States.* New York: E. J. Brill.

Herrick, G. W. 1913. *Control of Two Elm-Tree Pests.* Ithaca: Cornell University Agricultural Experiment Station Bulletin 333.

Hickman, J. C., ed. 1993. *The Jepson Manual: Higher Plants of California.* Berkeley: University of California Press.

Hoagland, R. E. 1990. *Microbes and Microbial Products as Herbicides.* Washington, D.C.: American Chemical Society Symposium Series 439.

Hodek, I. 1973. *Biology of Coccinellidae.* The Hague, Netherlands: Junk.

Hoddle, M. S., and R. Van Driesche. 1996. Evaluation of *Encarsia formosa* (Hymenoptera: Aphelinidae) to control *Bemisia argentifolii* (Homoptera: Aleyrodidae) on poinsettia (*Euphorbia pulcherrima*): A lifetable analysis. *Florida Entomologist* 79:1–12.

Hoffmann. M. P., C. H. Petzoldt, and A. C. Frodsham. 1996. *Integrated Pest Management for Onions.* Ithaca: Cornell University Cooperative Extension.

Hoffmann, M. P., L. T. Wilson, F. G. Zalom, and R. J. Hilton. 1991b. Dynamic sequential sampling plan for *Helicoverpa zea* (Lepidoptera: Noctuidae) eggs in processing tomatoes: parasitism and temporal patterns. *Environ. Entomol.* 20:1005–1012.

Hoffmann, M. P., L. T. Wilson, F. G. Zalom, R. J. Hilton, and C. V. Weakley. 1990. Parasitoid helps control fruitworm in Sacramento Valley processing tomatoes. *Calif. Agric.* 44(1):20–23.

Hoffmann, M. P., N. A. Davidson, L. T. Wilson, L. E. Ehler, W. A. Jones, and F. G. Zalom. 1991a. Imported wasp helps control southern green stink bug. *Calif. Agric.* 45(3):20–22.

Hoitink, H. A., and M. E. Grebus. 1994. Status of biological control of plant diseases with composts. *Compost Science Utilization* 2:6–12.

Hokkanen, H. M. T., and J. M. Lynch. 1995. *Biological Control: Benefits and Risks.* Cambridge: Cambridge University Press.

Hölldobler, B., and E. O. Wilson. 1990. *The Ants.* Cambridge, MA: Harvard University Press.

Holloway, J. K. 1957. Weed control by insect. *Scientific American* 197(1):56–62.

Hom, A. 1994. Current status of entomopathogenic nematodes. *IPM Practitioner* 16(3):1–12.

Honda, J. Y., and R. F. Luck. 1995. Scale morphology effects feeding behavior and biological control potential of *Rhyzobius lophanthae* (Coleoptera: Coccinellidae). *Ann. Entomol. Soc. Am.* 88:441–450.

Hoover, K., C. M. Schultz, S. S. Lane, B. C. Bonning, S. S. Duffey, B. F. McCutchen, and B. D. Hammock. 1995. Reduction in damage to cotton plants by a recombinant baculovirus that knocks moribund larvae of *Heliothis virescens* off the plant. *Biol. Control* 5:419–426.

Horn, D. J. 1996. Impacts of non-indigenous arthropods in biological control. *Midwest Biol. Control News* 3(12):1–2, 6.

Howarth, F. G. 1991. Environmental impacts of classical biological control. *Ann. Rev. Entomol.* 36:485–509.

Hoy, M. A. 1984. *Managing Mites in Almonds: An Integrated Approach.* Davis: Univ. Calif. Integrated Pest Management Publ. 1.

———. 1993. Transgenic beneficial arthropods for pest management programs: An assessment of their practicality and risks. In *Pest Management: Biologically Based Technologies*, R. D. Lumsden and J. L. Vaughn, eds. Conference Proceedings. Beltsville, MD: American Chemical Society. 357–369.

Hoy, M. A., W. W. Barnett, L. C. Hendricks, D. Castro, D. Cahn, and W. J. Bentley. 1984. Managing spider mites in almonds with pesticide-resistant predators. *Calif. Agric.* 38(7,8):18–20.

Hsieh, C. Y., and W. W. Allen. 1986. Effects of insecticides on emergence, survival, longevity, and fecundity of the parasitoid *Diaeretiella rapae* (Hymenoptera: Aphidiidae) from mummified *Myzus persicae* (Homoptera: Aphididae). *J. Econ Entomol.* 79:1599–1602.

Huang, H. T., and P. Yang. 1987. The ancient cultured citrus ant. *BioSci.* 37:665–671.

Huffaker, C. B., and C. E. Kennett. 1959. A ten-year study of vegetational changes associated with biological control of Klamath weed. *J. Range Mgmt.* 12:69–82.

Hull, L. A., and E. H. Beers. 1985. Ecological selectivity: Modifying chemical control practices to preserve natural enemies. In *Biological Control in Agricultural IPM Systems*, M. A. Hoy and D. C. Herzog, eds. New York: Academic Press. 103–122.

Hunter, C. D. 1997. Suppliers of Beneficial Organisms in North America. Sacramento, CA: Calif. Dept. of Pesticide Regulation.

Hussey, N. W., and N. Scopes, eds. 1985. *Biological Pest Control: The Glasshouse Experience.* Ithaca: Cornell University Press.

Idris, A. B., and E. Grafius. 1995. Wildflowers as nectar sources for *Diadegma insulare* (Hymenoptera: Ichneumonidae), a parasitoid of diamondback moth (Lepidoptera: Yponomeutidae). *Environ. Entomol.* 24 1726–1735.

———. 1996. Effects of wild and cultivated host plants on oviposition, survival, and development of diamondback moth (Lepidoptera: Plutellidae) and its parasitoid *Diadegma insulare* (Hymenoptera: Ichneumonidae). *Environ. Entomol.* 25:825–833.

Isakeit, T., S. Tjosvold, A. R. Weinhold, M. N. Schroth, and J. G. Hancock. 1991. Evaluation of several soil amendments added after fumigation with methyl bromide to control Fusarium wilt of carnation, 1989-90. *Biol. and Cult. Tests for Control of Plant Dis.* 6:105.

Isakeit, T., A. R. Weinhold, J. G. Hancock, M. N. Schroth, and S. Tjosvold. 1993. Counteraction of the biological vacuum created by soil fumigation and steaming. *Innovations in Landscape Pest Management,* Report of E. J. Slosson Fund for Ornamental Hort. Oakland: Univ. Calif. Div. Agric. Nat. Res. 46–48.

Jackson, D. M., and K. M. Kester. 1996. Effect of diet on longevity and fecundity of the spined stilt bug, *Jalysus wickhami*. *Entomologia Experimentalis et Applicata* 80:421–425.

Jaffee, B. A., and A. E. Muldoon. 1995. Susceptibility of root-knot and cyst nematodes to the nematode-trapping fungi *Monacrosporium ellipsosporum* and *M. cionopagum*. *Soil Biol. Biochem.* 27:1083–1090.

Jaffee, B. A., A. E. Muldoon, and E. C. Tedford. 1992. Trap production by nematophagous fungi growing from parasitized nematodes. *Phytopath.* 82:615–620.

Jaffee, B. A., A. E. Muldoon, and B. B. Westerdahl. 1996. Failure of mycelial formulation of the nematophagous fungus *Hirsutella rhossiliensis* to suppress the nematode *Heterodera schachtii*. *Biol. Control* 6:340–346.

Jarvis W. R. 1992. *Managing Diseases in Greenhouse Crops.* St. Paul, MN: APS Press.

Jepson. P. C., ed. 1989. *Pesticides and Non-target Invertebrates.* Wimborne, UK: Intercept.

Jones, S. A., and J. G. Morse. 1995. Use of isoelectric focusing electrophoresis to evaluate citrus thrips (Thysanoptera: Thripidae) predation by *Euseius tularensis* (Acari: Phytoseiidae). *Environ. Entomol.* 24:1040–1051.

Jones, V. P., M. P. Parrella, and D. R. Hodel. 1986. Biological control of leafminers in greenhouse chrysanthemums. *Calif. Agric.* 40(1/2):10–12.

Johnson, W. T., and H. H. Lyon. 1988. *Insects That Feed on Trees and Shrubs.* Ithaca: Cornell University Press.

Johnson, D. M., and P. D. Stiling. 1996. Host specificity of *Cactoblastis cactorum* (Lepidoptera: Pyralidae), an exotic *Opuntia*-feeding moth, in Florida. *Environ. Entomol.* 25:743–748.

Johnson, K. S., J. M. Scriber, J. K. Nitao, and D. R. Smitley. 1995. Toxicity of *Bacillus thuringiensis* var. *kurstaki* to three nontarget Lepidoptera in field studies. *Environ. Entomol.* 24:288–297.

Julien. M. H. 1992. *Biological Control of Weeds: A World Catalogue of Agents and Their Target Weeds.* 3d ed. Brisbane, Australia: CAB International.

Kaston, B. J. 1978. *How to Know the Spiders.* Dubuque, IA: Wm. C. Brown.

Kaya, H. K. 1993. Contemporary issues in biological control with entomopathogenic nematodes. Taipei City, Taiwan: Food and Fertilizer Technology Center Extension Bulletin 375.

Kaya, H. K., and R. Gaugler. 1993. Entomopathogenic nematodes. *Ann. Rev. Entomol.* 38:181–206.

Kaya, H. K., T. M. Burlando, H. Y. Choo, and G. S. Thurston. 1995. Integration of entomopathogenic nematodes with *Bacillus thuringiensis* or pesticidal soap for control of insect pests. *Biol. Control* 5:432–441.

Kelley-Tunis, K. K., B. L. Reid, and M. Andis. 1995. Activity of entomopathogenic fungi in free-foraging workers of *Camponotus pennsylvanicus* (Hymenoptera: Formicidae). *J. Econ. Entomol.* 88:937–943.

Kelton, L. A. 1978. *The Insects and Arachnids of Canada. Part 4: The Anthocoridae of Canada and Alaska (Heteroptera: Anthocoridae).* Ottawa: Canada Department of Agriculture Biosystematics Research Institute Publ. 1639.

———. 1982. *Plant Bugs on Fruit Crops in Canada (Heteroptera: Miridae).* Ottawa: Agriculture Canada Research Branch Biosystematics Research Institute Monograph 24.

Kevan, D. K. M., and J. Klimaszewski. 1987. The Hemerobiidae of Canada and Alaska Genus *Hemerobius*. *Giornale Italiano Di Entomologia* 3:305–369.

Kleintjes, P. K., and D. L. Dahlsten. 1994. Foraging behavior and nestling diet of chestnut-backed chickadees in Monterey pine. *The Condor* 96:647–653.

Knight, R. L., and M. K. Rust. 1991. Efficacy of formulated baits for control of Argentine ant (Hymenoptera: Formicidae). *J. Econ. Entomol.* 84:510–514.

Koppenhöfer, A. M., and H. K. Kaya. 1997. Additive and synergistic interaction between entomopathogenic nematodes and *Bacillus thuringiensis* for scarab grub control. *Biol. Control* 8:131–137.

Kouakou, B., D. Rampersad, E. Rodriguez, and D. L. Brown. 1992. Dairy goats used to clear poison oak do not transfer toxicant to milk. *Calif. Agric.* 46(3):4–6.

Kramer, V. L., R. Garcia, and A. E. Colwell. 1988. An evaluation of *Gambusia affinis* and *Bacillus thuringiensis* var. *israelensis* as mosquito control agents in California wild rice fields. *J. Am. Mosquito Control Assoc.* 4:470–478.

Krombein, K. V., P. D. Hurd, and D. R. Smith. 1979. *Catalog of Hymenoptera in America North of Mexico.* Washington, D.C.: Smithsonian Institution Press.

Laing, J. E., and J. Hamai. 1976. Biological control of insect pests and weeds by imported parasites, predators, and pathogens. In *Theory and Practice of Biological Control*, C. B. Huffaker and P. S. Messenger, eds. New York: Academic Press. 685–743.

LaMana, M. L., and J. C. Miller. 1996. Field observations on *Harmonia axyridis* Pallas (Coleoptera: Coccinellidae) in Oregon. *Biol. Control* 6:232–237.

Lampson, L. J., and J. G. Morse. 1992. A survey of black scale, *Saissetia oleae* (Hom.: Coccidae) parasitoids (Hym.: Chalcidoidea) in southern California. *Entomophaga* 37(3):373–390.

Lanini, W. T., J. Shribbs, and C. E. Elmore. 1988. Orchard Floor Mulching Trials in the U.S.A. *Le Fruit Belge* 56(3):228–249.

Lanini, W. T., C. D. Thomsen, T. S. Prather, C. E. Turner, J. M. DiTomaso, M. J. Smith, C. L. Elmore, M. P. Vayssieres, and W. A. Williams. 1995. *Yellow Starthistle.* Univ. Calf. Div. Agric. Nat. Res. Pest Notes 3.

Legner, E. F., and G. Gordh. 1992. Lower navel orangeworm (Lepidoptera: Phycitidae) population densities following establishment of *Goniozus legneri* (Hymenoptera: Bethylidae) in California. *J. Econ. Entomol.* 85:2153–2160.

Leigh, T. F., and P. B. Goodell. 1996. Insect management. In *Cotton Production Manual*, S. J. Hake, T. A. Kerby, and K. D. Hake, eds. Oakland: Univ. Calif. Div. Agric. Nat. Res. Publ. 3352. 260–293.

Levi, H. W., and L. R. Levi. 1990. *Spiders and Their Kin.* New York: Golden Press.

Lewis, J. A., R. D. Lumsden, and J. C. Locke. 1996. Biocontrol of damping-off diseases caused by *Rhizoctonia solani* and *Pythium ultimum* with alginate prills of *Gliocladium virens, Trichoderma hamatum* and various food bases. *Biocontrol Sci. Tech.* 6:163–173.

Lindow, S. E., G. McGourty, and R. Elkins. 1996. Interactions of antibiotics with *Pseudomonas fluorescens* strain A506 in the control of fire blight and frost injury to pear. *Phytopath.* 86:841–848.

Long, R. F. 1996. Bats for insect biocontrol in agriculture. *IPM Practitioner.* 18(9):1–6.

Losey, J. E., S. J. Fleischer, D. D. Calvin, W. L. Harkness, and T. Leahy. 1995. Evaluation of *Trichogramma nubilalis* and *Bacillus thuringiensis* in management of *Ostrinia nubilalis* (Lepidoptera: Pyralidae) in sweet corn. *Environ. Entomol.* 24:436–445.

Lövei, G. L., and K. D. Sunderland. 1996. Ecology and behavior of ground beetles (Coleoptera: Carabidae). *Ann. Rev. Entomol.* 41:231–256.

Luck, R. F. 1981. Parasitic insects introduced as biological control agents for arthropod pests. In *CRC Handbook of Pest Management in Agriculture.* Vol. 2, D. Pimentel, ed. Boca Raton, FL: CRC Press. 125–284.

Luck, R. F., and G. T. Scriven. 1976. The elm leaf beetle, *Pyrrhalta luteola*, in Southern California: Its pattern of increase and its control by introduced parasites. *Environ. Entomol.* 5:409–416.

Lumsden, R. D., and J. L. Vaughn, eds. 1993. *Pest Management: Biologically Based Technologies.* Conference Proceedings. Beltsville, MD: American Chemical Society.

Lumsden, R. D., J. A. Lewis, and J. C. Locke. 1993. Managing soilborne plant pathogens with fungal antagonists. In *Pest Management: Biologically Based Technologies.* R. D. Lumsden and J. L. Vaughn, eds. Conference Proceedings. Beltsville, MD: American Chemical Society. 196–203.

Mackauer, M., L. E. Ehler, and J. Roland, eds. 1990. *Critical Issues in Biological Control.* Andover, UK: Intercept.

Malais, M., and W. J. Ravensberg. 1992. *Knowing and Recognizing the Biology of Glasshouse Pests and Their Natural Enemies.* Berkel en Rodenrijs, Netherlands: Koppert Biological Systems.

Maredia, K. M., S. H. Gage, D. A. Landis, and J. M. Scriber. 1992. Habitat use patterns by the seven-spotted lady beetle (Coleoptera: Coccinellidae) in a diverse landscape. *Biol. Control* 2:159–165.

Marer, P. J., M. L. Flint, and M. W. Stimmann. 1988. *The Safe and Effective Use of Pesticides.* Oakland: Univ. Calif. Div. Agric. Nat. Res. Publ. 3324.

Markin, G. P., E. R. Yoshioka, and R. E. Brown. 1995. Gorse, *Ulex europaeus* L. Fabaceae. In *Biological Control in the Western United States.* J. R. Nechols et al., eds. Oakland: Univ. Calif. Div. Agric. Nat. Res. Publ. 3361.

Marois, J. J. 1992. Biological control of *Botrytis cinerea.* In *Biological Control of Plant Diseases*, E. S. Tjamos, G. C. Papavizas, and R. J. Cook, eds. New York: Plenum.

Marsh, P. M., S. R. Shaw, and R. A. Wharton. 1987. An identification manual for the North American genera of the family Braconidae (Hymenoptera). *Memoirs Entomol. Soc. Wash.* 13.

Maund, C. M., and T. H. Hsiao. 1991. Differential encapsulation of two *Bathyplectes* parasitoids among alfalfa weevil strains, *Hypera postica* (Gyllenhal). *Can. Ent.* 123:197–203.

Mays, W. T., and L. T. Kok. 1996. Establishment and dispersal of *Urophora affinis* (Diptera: Tephritidae) and *Metzneria paucipunctella* (Lepidoptera: Gelechiidae) in southwestern Virginia. *Biol. Control* 6:299–305.

McClintock, E., and A. T. Leiser. 1979. *An Annotated Checklist of Woody Ornamental Plants of California, Oregon, & Washington.* Oakland: Univ. Calif. Div. Agric. Nat. Res. Publ. 4091.

McEvoy, P., C. Cox, and E. Coombs. 1991. Successful biological control of ragwort, *Senecio jacobaea*, by introduced insects in Oregon. *Ecol. Appl.* 11:430–442.

McMurtry, J. A., and B. A. Croft. 1997. Life-history of phytoseiid mites and their roles in biological control. *Ann. Rev. Entomol.* 42:291–321.

McMurtry, J. A., H. G. Johnson, and S. J. Newberger. 1991. Imported parasite of greenhouse thrips established on California avocado. *Calif. Agric.* 45(6):31–32.

McMurtry, J. A., E. R. Oatman, P. A. Phillips, and C. W. Wood. 1978. Establishment of *Phytoseiulus persimilis* (Acari: Phytoseiidae) in Southern California. *Entomophaga* 23:175–179.

Mensah, R. K., and J. L. Madden. 1993. Development and application of an integrated pest management programme for the psyllid, *Ctenarytaina thysanura* on *Boronia megastigma* in Tasmania. *Entomologia Experimentalis et Applicata* 66:59–74.

Merlin, J., O. Lemaitre, and J. C. Grégoire. 1996. Oviposition in *Cryptolaemus montrouzieri* stimulated by wax filaments of their prey. *Entomologia Experimentalis et Applicata* 79:141–146.

Messing, R. H., and M. T. Aliniazee. 1989. Introduction and establishment of *Trioxys pallidus* (Hym.: Aphidiidae) in Oregon, U.S.A. for control of filbert aphid *Myzocallis coryli* (Hom.: Aphidiidae). *Entomophaga* 34:153–163.

Meyer, S. L., and R. N. Huettel 1993. Fungi and fungus/bioregulator combinations for control of plant-parasitic nematodes. In *Pest Management: Biologically Based Technologies*, R. D. Lumsden and J. L. Vaughn, eds. Conference Proceedings. Beltsville, MD: American Chemical Society. 214–221.

Meyerdirk, D. E., I. M. Newell, and R. W. Warkentin. 1981. Biological control of Comstock mealybug. *J. Econ. Entomol.* 74:79–84.

Miklasiewicz, T. J., and G. P. Walker. 1990. Population dynamics and biological control of the woolly whitefly (Homoptera: Aleyrodidae) on citrus. *Environ. Entomol.* 19:1485–1490.

Miller, P. R., W. L. Graves, W. A. Williams, and B. A. Madison. 1989. *Covercrops for California Agriculture.* Oakland: Univ. Calif. Div. Agric. Nat. Res. Publ. 21471.

Moreno, D. S., and R. F. Luck. 1992. Augmentative releases of *Aphytis melinus* (Hymenoptera: Aphelinidae) to suppress California red scale (Homoptera: Diaspididae) in Southern California lemon orchards. *J. Econ. Entomol.* 85:1112–1119.

Murphy, B. C., J. A. Rosenheim, and J. Granett. 1996. Habitat diversification for improving biological control: Abundance of *Anagrus epos* (Hymenoptera: Mymaridae) in grape vineyards. *Environ. Entomol.* 25:495–504.

Nechols, J. R., L. A. Andres, J. W. Beardsley, R. D. Goeden, and C. G. Jackson, eds. 1995. *Biological Control in the Western United States: Accomplishments and Benefits of Regional Project W-84, 1964–1989.* Oakland: Univ. Calif. Div. Agric. Nat. Res. Publ. 3361.

Nordlund, D. A., D. C. Vacek, and D. N. Ferro. 1991. Predation of Colorado potato beetle (Coleoptera: Chrysomelidae) eggs and larvae by *Chrysoperla rufilabris* (Neuroptera: Chrysopidae) in the laboratory and field cages. *J. Entomol. Sci.* 26:443–449.

Oatman, E. R., and G. R. Platner. 1985. Biological control of two avocado pests. *Calif. Agric.* 39(11/12):21–23.

Oatman, E. R., J. A. Wyman, R. A. van Steenwyk, and M. W. Johnson. 1983. Integrated control of tomato fruitworm (Lepidoptera: Noctuidae) and other lepidopterous pests on fresh-market tomatoes in Southern California. *J. Econ. Entomol.* 76:1363–1369.

Obrycki, J. J. 1989. Parasitization of native and exotic coccinellids by *Dinocampus coccinellae* (Schrank) (Hymenoptera: Braconidae). *J. Kansas Entomol. Soc.* 62:211–218.

Olkowski, W., S. Daar, and H. Olkowski. 1991. *Common-Sense Pest Control.* Newtown, CT: Taunton Press.

O'Neil, B. 1997. You get what you pay for: Quality control of natural enemies. *Midwest Biol. Control News* 4(1):1–2, 6–7.

O'Neill, T. M., Y. Elad, D. Shtienberg, and A. Cohen. 1996. Control of grapevine grey mould with *Trichoderma harzianum* T39. *Biocontrol Sci. Tech.* 6:139–146.

Osborne, L. S., and L. E. Ehler. 1981. Biological control of twospotted spider mite in California greenhouses. Oakland: Univ. Calif. Div. Agric. Nat. Res. Leaflet 21271.

Osborne, L. S., L. E. Ehler, and J. R. Nechols. 1985. *Biological Control of Twospotted Spider Mite in Greenhouses.* Gainesville: Univ. Florida Agric. Exp. Sta. Bull. 853.

Ouyang, Y., E. E. Grafton-Cardwell, and R. L. Bugg. 1992. Effects of various pollens on development, survivorship, and reproduction of *Euseius tularensis* (Acari: Phytoseiidae). *Environ. Entomol.* 21:1371–1376.

Packard, A. S. 1876. *Guide to the Study of Insects and a Treatise on those Injurious and Beneficial to Crops.* New York: Henry Holt.

Papp, C. S. 1984. *Introduction to North American Beetles.* Sacramento, CA: Entomography Publications.

Parrella, M. P., T. D. Paine, J. A. Bethke, K. L. Robb, and J. Hall. 1991. Evaluation of *Encarsia formosa* for biological control of sweetpotato whitefly (Homoptera: Aleyrodidae) on poinsettia. *Environ. Entomol.* 20:713–719.

Pemberton, R. W. 1995a. *Cactoblastis cactorum* (Lepidoptera: Pyralidae) in the United States: An immigrant biological control agent or an introduction of the nursery industry. *Am. Entomologist* 41:230–232.

———. 1995b. Leafy spurge, *Euphorbia esula* L. Euphorbiaceae. In *Biological Control in the Western United States.* J. R. Nechols et al., eds. Oakland: Univ. Calif. Div. Agric. Nat. Res. Publ. 3361. 289–295.

Pemberton, R. W., and C. E. Turner. 1990. Biological control of *Senecio jacobaea* in Northern California, an enduring success. *Entomophaga* 35:71–77.

Peterson, A. 1960. *Larvae of Insects.* Part 2. Ann Arbor, MI: Edwards Brothers.

Pfleger, F. L., and R. G. Linderman. 1994. *Mycorrhizae and Plant Health.* St. Paul, MN: APS Press.

Phillips, P. A., and C. J. Sherk. 1991. To control mealybugs, stop honeydew-seeking ants. *Calif. Agric.* 45(2):26–28.

Phillips, P. A., R. S. Bekey, and G. E. Goodall. 1987. Argentine ant management in cherimoyas. *Calif. Agric.* 41(3,4):8–9.

Phoofolo, M. W., and J. J. Obrycki. 1995. Comparative life-history studies of Nearctic and Palearctic populations of *Coccinella septempunctata* (Coleoptera: Coccinellidae). *Environ. Entomol.* 24:581–587.

Pickel, C., P. Phillips, J. Trumble, N. Welch, and F. Zalom. 1996. Strawberry Pest Management Guidelines: Insects and Mites. UC IPM Guidelines Series 22. In *UC IPM Pest Management Guidelines*, Univ. Calif. Div. Agric. Nat. Res. Publ. 3339.

Pickett, C. H., S. E. Schoenig, and M. P. Hoffmann. 1996. Establishment of the squash bug parasitoid, *Trichopoda pennipes* Fabr. (Diptera: Tachinidae), in Northern California. *Pan-Pac. Entomologist* 72:220–226.

Pickett, C. H., J. C. Ball, K. C. Casanave, K. M. Klonsky, K. M. Jetter, L. G. Bezark, and S. E. Schoenig. 1996. Establishment of the ash whitefly parasitoid *Encarsia inaron* (Walker) and its economic benefit to ornamental street trees in California. *Biol. Control* 6:260–272.

Piper, G. L., and L. A. Andres. 1995. Canadian thistle, *Cirsium arvense* (L.) Scop. Asteraceae. In *Biological Control in the Western United States.* J. R. Nechols et al., eds. Oakland: Univ. Calif. Div. Agric. Nat. Res. Publ. 3361. 233–236.

Piper, G. L., and S. S. Rosenthal. 1995. Diffuse knapweed, *Centaurea diffusa* Lamarck Asteraceae. In *Biological Control in the Western United States.* J. R. Nechols et al., eds. Oakland: Univ. Calif. Div. Agric. Nat. Res. Publ. 3361. 237–241.

Powell, J. E., and C. L. Hogue 1979. *California Insects.* Berkeley: University of California Press.

Provencher, L., and S. E. Riechert. 1994. Model and field test of prey control effects by spider assemblages. *Environ. Entomol.* 23:1–17.

Quarles, W. 1996a. Stored product biocontrol. *IPM Practitioner* 18(5/6):13–17.

———. 1996b. New microbial pesticides for IPM. *IPM Practitioner* 18(8):5–10.

Quarles, W., and J. Grossman. 1995. Alternatives to methyl bromide in nurseries—Disease suppressive media. *IPM Practitioner* 17(8):1–13.

Quayle, H. J. 1932. *Biology and Control of Citrus Insects and Mites.* Berkeley: Univ. Calif. Agric. Exper. Sta. Bull. 542.

Quezada, J. R., and P. DeBach. 1973. Bioecological and population studies of the cottonycushion scale, *Icerya purchasi* Mask., and its natural enemies, *Rodolia cardinalis* Mul. and *Cryptochaetum iceryae* Will., in Southern California. *Hilgardia* 41:631–688.

Raupp, M. J., M. R. Hardin, S. M. Braxton, and B. B. Bull. 1994. Augmentative releases for aphid control on landscape plants. *J. Arbor.* 20:241–249.

Rice, E. L. 1995. *Biological Control of Weeds and Plant Diseases: Advances in Applied Allelopathy.* Norman: University of Oklahoma Press.

Riddick, E. W., and N. J. Mills. 1994. Potential of adult carabids (Coleoptera: Carabidae) as predators of fifth-instar codling moth (Lepidoptera: Tortricidae) in apple orchards in California. *Environ. Entomol.* 23:1338–1345.

Ridgway, R. L., and W. L. Murphy. 1984. Biological control in the greenhouse. In *Biology of Chrysopidae*, T. R. New, ed. The Hague, Netherlands: Junk. 220–228.

Riechert, S. E., and L. Bishop. 1990. Prey control by an assemblage of generalist predators: Spiders in garden test systems. *Ecology* 71:1441–1450.

Rose, M., and P. DeBach. 1982. A native parasite of the bayberry whitefly. *Citrograph* 67:272–276.

Rose, M., and G. Zolnerowich. 1997. *The Genus Eretmocerus (Hymenoptera: Aphelinidae) Parasites of Whitefly (Homoptera: Aleyrodidae).* College Station: Texas A&M University, Department of Entomology.

Rosenheim, J. A., and M. A. Hoy. 1988. Sublethal effects of pesticides on the parasitoid *Aphytis melinus* (Hymenoptera: Aphelinidae). *J. Econ. Entomol.* 81:476–483.

Rosenheim, J. A., and L. R. Wilhoit. 1993. Why lacewings may fail to suppress aphids: Predators that eat other predators disrupt cotton aphid control. *Calif. Agric.* 47(5):7–9.

Roush, R. T., and M. A. Hoy. 1981. Laboratory, glasshouse, and field studies of artificially selected carbaryl resistance in *Metaseiulus occidentalis*. *J. Econ. Entomol.* 74:142–147.

Sakovich, N. J., J. B. Bailey, and T. W. Fisher. 1984. *Decollate Snails for Control of Brown Garden Snails in Southern California Citrus Groves.* Oakland: Univ. Calif. Div. Agric. Nat. Res. Leaflet 21384.

Sanderson, E. D., and C. F. Jackson. 1912. *Elementary Entomology.* Boston: Ginn.

Santha, C. R., R. D. Martyn, W. H. Neill, and K. Strawn. 1994. Control of submersed weeds by grass carp in waterlily production ponds. *J. Aquat. Plant Mgmt.* 32:29–33.

Schauff, M. E., G. A. Evans, and J. M. Heraty. 1996. A pictorial guide to the species of *Encarsia* (Hymenoptera: Aphelinidae) parasitic on whiteflies (Homoptera: Aleyrodidae) in North America. *Proc. Entomol. Soc. Wash.* 98:1–35.

Schoenig, S. E., and C. H. Pickett. 1996. Establishment and distribution of the tachinid fly, *Trichopoda pennipes*, for the biological control of the squash bug, *Anasa tristis*. In *Biological Control Program Annual Summary*, L. G. Bezark, ed. Sacramento: California Department of Food and Agriculture, Division of Plant Industry.

Schroth, M. N., and J. G. Hancock. 1982. Disease-suppressive soil and root colonizing bacteria. *Science* 216:1379–1381.

Schroth, M. N., J. G. Hancock, and A. R. Weinhold. 1992. Biological approaches to control of plant diseases. In *Beyond Pesticides: Biological Approaches to Pest Management in California*. Oakland: Univ. Calif. Div. Agric. Nat. Res. Publ. 3354. 102–122.

Schuh, R. T., and J. A. Slater. 1995. *True Bugs of the World (Hemiptera: Heteroptera): Classification and Natural History.* Ithaca: Cornell University Press.

Schweizer, H., and J. G. Morse. 1997. Factors influencing survival of citrus thrips (Thysanoptera: Thripidae) propupae and pupae on the ground. *J. Econ. Entomol.* 90:435–443.

Shattuck, S. O. 1985. *Illustrated Key to Ants Associated with Western Spruce Budworm.* Washington, D.C.: U.S. Dept. Agric. Handb. 632.

Shorey, H. H., L. K. Gaston, R. G. Gerber, C. B. Sisk, and P. A. Phillips. 1996. Formulating farnesol and other ant-repellent semiochemicals for exclusion of Argentine ants (Hymenoptera: Formicidae) from citrus trees. *Environ. Entomol.* 25:114–119.

Sikora, R. A., and S. Hoffmann-Hergarten. 1993. Biological control of plant parasitic nematodes with plant-health-promoting rhizobacteria. In *Pest Management: Biologically Based Technologies*, R. D. Lumsden and J. L. Vaughn, eds. Conference Proceedings. Beltsville, MD: American Chemical Society. 166–172.

Silvestri, F. 1910. Contribuzioni alla conoscenza degli insetti dannosi e dei loro simbionti. *Bollettino del Laboratorio di Zoologia Generale e Agraria* (Portici) 4:246–288.

Simanton, F. L. 1916. *Hyperaspis binotata*, a predatory enemy of the terrapin scale. *J. Agric. Res.* 6:197–204.

Sisk, C. B., H. H. Shorey, R. G. Gerber, and L. K. Gaston. 1996. Semiochemicals that disrupt foraging by the Argentine ant (Hymenoptera: Formicidae): Laboratory bioassays. *J. Econ. Entomol.* 89:381–385.

Slater, J. A., and R. M. Baranowski. 1978. *How to Know the True Bugs.* Dubuque, IA: Wm. C. Brown.

Smirnoff, W. A. 1959. Predators of *Neodiprion swainei* Midd. (Hymenoptera: Tenthredinidae), larval vectors of virus diseases. *Can. Ent.* 91:246–248.

Smith, K. M., and A. W. Cressman. 1962. Birefringent crystals in virus-infected citrus red mites. *J. Insect Path.* 4:229–236.

Smith, M. W., D. C. Arnold, R. D. Eikenbary, N. R. Rice, A. Shiferaw, B. S. Cheary, and B. L. Carroll. 1996. Influence of ground cover on beneficial arthropods in pecans. *Biol. Control* 6:164–176.

Smith, R. F., and K. S. Hagen. 1956. Enemies of spotted alfalfa aphid. *Calif. Agric.* 10(4):8–10.

Smith, R. J. 1986. Biological control of northern jointvetch (*Aeschynomene virginica*) in rice (*Oryza sativa*) and soybeans (*Glycine max*)—A researcher's view. *Weed Sci.* 34 (Suppl. 1):17–23.

Smith, S. M. 1996. Biological control with *Trichogramma*: Advances, success, and potential of their use. *Ann. Rev. Entomol.* 41:375–406.

Spollen, K. M., and M. A. Hoy. 1992. Genetic improvement of an arthropod natural enemy. *Biol. Control* 2:87–94.

Spollen, K. M., M. W. Johnson, and B. E. Tabashnik. 1995. Stability of fenvalerate resistance in the leafminer parasitoid *Diglyphus begini* (Hymenoptera: Eulophidae). *J. Econ. Entomol.* 88:192–197.

Stapleton, J. J., and J. E. DeVay. 1984. Thermal components of soil solarization as related to changes in soil and root microflora and increased plant growth response. *Phytopath.* 74:255–259.

———. 1995. Soil solarization: A natural mechanism of integrated pest management. In *Novel Approaches to Integrated Pest Management*, R. Reuveni, ed. Boca Raton, FL: Lewis Pub.

Starnes, R. L., C. L. Liu, and P. G. Marrone. 1993. History, use, and future of microbial insecticides. *Am. Entomologist* 39:83–91.

Starý, P. 1991. *Philadelphus coronarius* L. as a reservoir of aphids and parasitoids. *J. Appl. Ent.* 112:1–10.

———. 1993. Alternative host and parasitoid in first method in aphid pest management in glasshouses. *J. Appl. Ent.* 16:187–191.

Steffan, S., and P. Whitaker. 1996. Guarding the garden: habitat manipulation to favor natural enemies. *Midwest Biol. Control News* 3(4):1–2, 7.

Stehr, F. W. 1991. *Immature Insects.* Vol. 2. Dubuque, IA: Kendall Hunt.

Sterling, G. R. 1991. *Biological Control of Plant Parasitic Nematodes: Progress, Problems, and Prospects.* Wallingford, UK: CAB International.

Sterling, W. L., D. A. Dean, D. A. Fillman, and D. Jones. 1984. Naturally-occurring biological control of the boll weevil (Col.: Curculionidae). *Entomophaga* 29:1–9.

Stern, V. M. 1969. Interplanting alfalfa in cotton to control lygus bug and other insect pests. *Proc. Tall Timbers Conference on Ecological Animal Control by Habitat Management* 1:55–69.

Stern, V. M., R. F. Smith, R. van den Bosch, and K. S. Hagen. 1959. The integrated control concept. *Hilgardia* 29:81–101.

Stiling, P. 1990. Calculating establishment rates of parasitoids in classical biological control. *Am. Entomologist* 36:225–230.

Stoetzel, M. B. 1989. *Common Names of Insects & Related Organisms 1989.* Lanham, MD: Entomological Society of America.

Story, J. M. 1995. Spotted knapweed, *Centaurea maculosa* Lamarck Asteraceae. In *Biological Control in the Western United States*. J. R. Nechols et al., eds. Oakland: Univ. Calif. Div. Agric. Nat. Res. Publ. 3361. 258–263.

Strand, L. L. 1994. *Integrated Pest Management for Strawberries.* Oakland: Univ. Calif. Div. Agric. Nat. Res. Publ. 3351.

Strandberg, J. O. 1981. Predation of cabbage looper, *Trichoplusia ni*, pupae by the striped earwig, *Labidura riparia*, and two bird species. *Environ. Entomol.* 10:712–715.

Summers, C. G. 1976. Population fluctuations of selected arthropods in alfalfa: Influence of two harvesting practices. *Environ. Entomol.* 5:103–110.

Summers, C. G., K. S. Hagen, and V. M. Stern. 1996. Alfalfa Pest Management Guidelines: Insects and Mites. UC IPM Guidelines Series 2. In *UC IPM Pest Management Guidelines*, Univ. Calif. Div. Agric. Nat. Res. Publ. 3339.

Sunderland, K. D., and G. P. Vickerman. 1980. Aphid feeding by some polyphagous predators in relation to aphid density in cereal fields. *J. Applied Ecology* 17:389–396.

Supkoff, D. M., D. B. Joley, and J. J. Marois. 1988. Effect of introduced biological control organisms on the density of *Chondrilla juncea* in California. *J. Applied Ecology* 25:1089–1095.

Swadling, I. R., and P. Jeffries. 1996. Isolation of microbial antagonists for biocontrol of grey mould disease of strawberries. *Biocontrol Sci. Tech.* 6:125–136.

Tamaki, G. 1981. Biological control of potato pests. In *Potato Pest Management*, J. H. Lashomb and R. Casagrande, eds. Stroudsburg, PA: Hutchinson Ross. 178–191.

Tamaki, G., and G. E. Long. 1978. Predator complex of the green peach aphid on sugarbeets: Expansion of the predator power and efficacy model. *Environ. Entomol.* 7:835–842.

Tanigoshi, L. K., and J. A. McMurtry. 1977. The dynamics of predation of *Stethorus picipes* (Coleoptera: Coccinellidae) and *Typhlodromus floridanus* on the prey *Oligonychus punicae* (Acarina: Phytoseiidae, Tetranychidae). *Hilgardia* 45:237–288.

Tassan, R. L., and K. S. Hagen. 1970. Culturing green lacewings in the home and school. Oakland: Univ. Calif. Div. Agric. Sci. Leaflet 2500.

Tauber, C. A. 1974. Systematics of North American chrysopid larvae: *Chrysopa carnea* group (Neuroptera). *Can. Ent.* 106:1133–1153.

———. 1991. Order Neuroptera. In *Immature Insects*. Vol. 2, F. W. Stehr, ed. Dubuque, IA: Kendall Hunt. 126–143.

Tedford, E. C., B. A. Jaffee, and A. E. Muldoon. 1995. Suppression of the nematode *Heterodera schachtii* by the fungus *Hirsutella rhossiliensis* as affected by fungus population density and nematode movement. *Phytopath.* 85:613–617.

Theiling, K. M., and B. A. Croft. 1988. Pesticide side-effects on arthropod natural enemies: A database summary. *Agric. Ecosystems Environ.* 21:191–218.

Thomsen, C. D., W. A. Williams, M. P. Vayssières, C. E. Turner, and W. T. Lanini. 1996. *Yellow Starthistle: Biology and Control*. Oakland: Univ. Calif. Div. Agric. Nat. Res. Leaflet 21541.

Thomson, S. V., D. R. Hansen, K. M. Flint, and J. D. Vandenberg. 1992. Dissemination of bacteria antagonistic to *Erwinia amylovora* by honey bees. *Plant Dis.* 76:1052–1056.

Tilghman, N. G. 1987. Characteristics of urban woodlands affecting breeding bird diversity and abundance. *Lands. Urban Plan.* 14:481–495.

Torgersen, T. R., and A. S. Torgersen. 1995. *Save Our Birds—Save Our Forests*. Portland, OR: U.S. Dept. Agric. For. Serv. Pacific Northwest Research Station.

Torgersen, T. R., J. W. Thomas, R. R. Mason, and D. V. Horn. 1984. Avian predators of Douglas-fir tussock moth, *Orgyia pseudotsugata* (McDunnough) (Lepidoptera: Lymantriidae) in southwestern Oregon. *Environ. Entomol.* 13:1018–1022.

Toscano, N. C., F. G. Zalom, and J. T. Trumble. 1995. Tomato Pest Management Guidelines: Insects and Mites. UC IPM Guidelines Series 14. In *UC IPM Pest Management Guidelines*, Univ. Calif. Div. Agric. Nat. Res. Publ. 3339.

Townes, H., and M. Townes. 1960. The ichneumon-flies of America north of Mexico. Part 2. *U.S. Nat. Museum Bull.* 216.

Triapitsyn, S. V., and D. H. Headrick. 1995. A review of the Nearctic species of the thrips-attacking genus *Ceranisus* Walker (Hymenoptera: Eulophidae). *Trans. Am. Entomol. Soc.* 121:227–248.

Trumble, J. T., and J. G. Morse. 1993. Economics of integrating the predaceous mite *Phytoseiulus persimilis* (Acari: Phytoseiidae) with pesticides in strawberries. *J. Econ. Entomol.* 86:879–885.

Trumble, J. T., W. G. Carson, and K. K. White. 1994. Economic analysis of a *Bacillus thuringiensis*-based IPM program in fresh-market tomatoes. *J. Econ. Entomol.* 87:1463–1469.

Tsai, J. H., and B. Steinberg. 1991. Current status of the citrus blackfly, *Aleurocanthus woglumi* (Homoptera: Aleyrodidae), and its parasites in south Florida. *Florida Entomologist* 74:153–156.

Tulisalo, U. 1984. Biological control in the greenhouse. In *Biology of Chrysopidae*. T. R. New., ed. The Hague, Netherlands: Junk. 229–233.

Turner, C. E., J. B. Johnson, and J. P. McCaffrey. 1995. Yellow starthistle, *Centaurea solstitialis* L. Asteraceae. In *Biological Control in the Western United States*. J. R. Nechols et al., eds. Oakland: Univ. Calif. Div. Agric. Nat. Res. Publ. 3361. 270–275.

Turner, C. E., L. W. Anderson, P. Foley, R. D. Goeden, W. T. Lanini, S. E. Lindow, and C. O. Qualset. 1992. Biological approaches to weed management. In *Beyond Pesticides: Biological Approaches to Pest Management in California*. Oakland: Univ. Calif. Div. Agric. Nat. Res. Publ. 3354. 32–67.

Tuzun, S., and J. Kloepper. 1995. Practical application and implementation of induced resistance. In *Induced Resistance to Disease in Plants*, R. Hammerschmidt and J. Kuć, eds. Dordrecht, Netherlands: Kluwer Academic. 152–168.

Umoru, P. A., W. Powell, and S. J. Clark. 1996. Effect of primicarb on the foraging behavior of *Diaeretiella rapae* (Hymenoptera: Braconidae) on host-free and infested oilseed rape plants. *Bull. Entomol. Res.* 86:193–201.

Usinger, R. L. 1971. *Aquatic Insects of California*. Berkeley: University of California Press.

van den Bosch, R., and K. S. Hagen. 1966. *Predaceous and Parasitic Arthropods in California Cotton Fields*. Berkeley: Univ. Calif. Agric. Exp. Sta. Bull. 820.

van den Bosch, R., P. S. Messenger, and A. P. Gutierrez. 1982. *An Introduction to Biological Control*. New York: Plenum.

van den Bosch, R., R. Hom, P. Matteson, B. D. Frazer, P. S. Messenger, and C. S. Davis. 1979. Biological control of the walnut aphid in California; impact of the parasite, *Trioxys pallidus*. *Hilgardia* 47:1–13.

van den Meiracker, R. A. F. 1994. Induction and termination of diapause in *Orius* predatory bugs. *Entomologia Experimentalis et Applicata* 73:127–137.

Vander Meer, R. K., K. Jaffee, and A. Cedeno. 1990. *Applied Myrmecology: A World Perspective*. Boulder, CO: Westview Press.

Van Driesche, R. G., and T. S. Bellows. 1993. *Steps in Classical Arthropod Biological Control*. Lanham, MD: Entomological Society of America.

———. 1996. *Biological Control*. New York: Chapman & Hall.

Van Driesche, R. G., S. Healy, and R. C. Reardon. 1996. *Biological Control of Arthropod Pests of the Northeastern and North Central Forests in the United States: A Review and Recommendations*. Morgantown, WV: U.S. Dept. Agric. Forest Serv. FHTET-96-19.

van Horn, M. 1995. *Compost Production and Utilization: A Growers' Guide*. Oakland: Univ. Calif. Div. Agric. Nat. Res. Leaflet 21514.

van Houten, Y. M., P. C. J. van Rijn, L. K. Tanigoshi, P. van Stratum, and J. Bruin. 1995. Preselection of predatory mites to improve year-round biological control of western flower thrips in greenhouse crops. *Entomologia Experimentalis et Applicata* 74:225–234.

van Lenteren, J. C., and A. J. M. Loomans. 1995. Biological control of thrips pests. *Wageningen* (Netherlands) *Agricultural University Papers* 95-1.

van Lenteren, J. C., and J. Woets. 1988. Biological and integrated pest control in greenhouses. *Ann. Rev. Entomol.* 33:239–269.

Vasconcelos, S. D., T. Williams, R. S. Hails, and J. S. Cory. 1996. Prey selection and baculovirus dissemination by carabid predators of Lepidoptera. *Ecological Ent.* 21:98–104.

Vaughn, T. T., M. F. Antolin, and L. B. Bjostad. 1996. Behavioral and physiological responses of *Diaeretiella rapae* to semiochemicals. *Entomologia Experimentalis et Applicata* 78:187–196.

Vockeroth, J. R. 1992. *The Insects and Arachnids of Canada. Part 18: The Flower Flies of the Subfamily Syrphinae of Canada, Alaska, and Greenland (Diptera: Syrphidae)*. Ottawa: Research Branch, Agriculture Canada Publ. 1867.

Wajnberg, E., and S. A. Hassan. 1994. *Biological Control with Egg Parasitoids*. Berkshire, UK: CAB International.

Walker, G. P., and D. C. G. Aitken. 1993. Development time of *Parabemisia myricae* (Kuwana) (Hom., Aleyrodidae) in different seasons and a degree-day model. *J. Appl. Ent.* 398–404.

Watkins, G. M., ed. 1981. *Compendium of Cotton Diseases*. St. Paul, MN: American Phytopathological Society.

Way, M. J., and K. C. Khoo. 1992. Role of ants in pest management. *Ann. Rev. Entomol.* 37:479–503.

Weber, C. A., J. M. Smilanick, L. E. Ehler, and F. G. Zalom. 1996. Ovipositional behavior and host discrimination in three scelionid egg parasitoids of stink bugs. *Biol. Control* 6:245–252.

Weseloh, R. M. 1994. Forest ant (Hymenoptera: Formicidae) effect on gypsy moth (Lepidoptera: Lymantriidae) larval numbers in a mature forest. *Environ. Entomol.* 23:870–877.

———. 1995. Forest characteristics associated with abundance of foraging ants (Hymenoptera: Formicidae) in Connecticut. *Environ. Entomol.* 24:1453–1457.

Wheeler, A. G., and T. J. Henry. 1981. *Jalysus spinosus* and *J. wickhami*: Taxonomic clarification, review of host plants and distribution, and keys to adults and 5th instars. *Ann. Entomol. Soc. Am.* 74:606–615.

Whipps, J. M. 1992. Status of biological disease control in horticulture. *Biocontrol Sci. Tech.* 2:3–24.

Wildermuth, V. L. 1914. *The Alfalfa Caterpillar.* Washington, D.C.: U.S. Dept. Agric. Bull. 124.

Wilson, F. 1943. *The Entomological Control of St. John's Wort (Hypericum perforatum L.).* Melbourne: Council for Scientific and Industrial Res. Bull. 169.

Wright, D. J., and R. H. J. Verkerk. 1995. Integration of chemical and biological control systems for arthropods: Evaluation in a multitrophic context. *Pesticide Sci.* 44:207–218.

Wright, E. M., and R. J. Chambers. 1994. The biology of the predatory mite *Hypoaspis miles* (Acari: Laclapidae), a potential biological control agent of *Bradysia paupera* (Dipt.: Sciaridae). *Entomophaga* 39:225–235.

Wurtz, T. L. 1995. Domestic geese: Biological weed control in an agricultural setting. *Ecol. Applic.* 5:570–578.

Zak, B., and I. Ho. 1994. Resistance of ectomycorrhizal fungi to Rhizinia root rot. *Indian J. Mycol. Pl. Path.* 24:192–195.

Zalom, F. G., R. A. Van Steenwyk, W. J. Bentley, R. Coviello, R. E. Rice, W. W. Barnett, C. Pickel, and M. M. Barnes. 1996. Almond Pest Management Guidelines: Insects. UC IPM Guidelines Series 1. In *UC IPM Pest Management Guidelines*, Univ. Calif. Div. Agric. Nat. Res. Publ. 3339.

Zchori-Fein, E., R. T. Roush, and J. P. Sanderson. 1994. Potential for integration of biological and chemical control of greenhouse whitefly (Homoptera: Aleyrodidae) using *Encarsia formosa* (Hymenoptera: Aphelinidae) and abamectin. *Environ. Entomol.* 23:1277–1282.

Zhang, J., C. R. Howell, and J. L. Starr. 1996. Suppression of *Fusarium* colonization of cotton roots and Fusarium wilt by seed treatments with *Gliocladium virens* and *Bacillus subtilis. Biocontrol Sci. Tech.* 6:175–187.

Zhang, Z., and J. P. Sanderson. 1995. Twospotted spider mite (Acari: Tetranychidae) and *Phytoseiulus persimilis* (Acari: Phytoseiidae) on greenhouse roses: Spatial distribution and predator efficacy. *J. Econ. Entomol.* 88:352–357.

Zheng, Y. n.d. *Head Markings of Green Lacewings Found in the San Joaquin Valley Vineyards.* University of California. unpublished

Zimmerman, E. C. 1948. *Insects of Hawaii.* Volume 3: *Heteroptera.* Honolulu: University of Hawaii Press.

Names of Pests and Plants

BOTH COMMON AND SCIENTIFIC names are used to identify organisms. Because different humans (*Homo sapiens*) may use different names for the same organism, names are often a source of confusion.

Scientists use a unique, two-word combination for each animal, plant, and microorganism. This scientific name provides the surest identification because scientific names are used according to agreed-upon rules. Although scientific names are sometimes changed based on new information, each organism has only one valid scientific name, which is used throughout the world. If a scientific name has recently been changed, both names may be printed: *Cotesia* (=*Apanteles*), with the currently correct name listed first followed by an equal sign and the former name in parentheses.

The first word of a scientific name, the genus or generic name, is capitalized. The second word, the species or specific name, is not. Both words are italicized and are also Latinized so scientists can understand what plant or animal others are referring to, regardless of nationality and native language. After its first use in the text, the genus name is often abbreviated; for example, the red scale parasite *Aphytis melinus* is shortened to *A. melinus.* When several species within the same genus are discussed together, species may be abbreviated as "spp." (e.g., *Aphytis* spp.). When referring to only one species, "sp." is used. A third scientific name, subspecies (abbreviated spp.), may also be used: *Bacillus thuringiensis* subspecies *tenebrionis.* Subspecies are especially common among microorganisms, which are more difficult to categorize than larger organisms.

Scientific names are used in a hierarchical organization or ranked order that includes the family names (family) listed in the Index. These hierarchical names show relationships among organisms, as illustrated here for the common convergent lady beetle:

Kingdom: Animalia (animals)
Class: Insecta (insects)
Order: Coleoptera (beetles)
Family: Coccinellidae
(lady beetles)
Genus: *Hippodamia*
Species: *convergens*

Besides the two-part scientific name, many plants, insects, and diseases also have common names. Common names are familiar to more people than scientific names, and they are often easier to pronounce and remember. However, there are serious problems with common names. There are no clear rules for deciding what is the correct common name of most organisms. A single common name is often used to refer to several distinctly different organisms. The same organism can have several common names, some of which may be known and used only by people in certain locations. For example, what is commonly known in California as avocado is in some parts of the southern United States called alligator-pear! Common names may also be ridiculous or inaccurate; pineapple refers to a plant that is very unlike pines and apples. Ladybug refers to certain beetles, which are very different from the insects that scientists call true bugs. Many important organisms, including most species of beneficial predators, parasites, and pathogens, have no common name because they are often tiny and known only to scientists.

The scientific name and one or more common names of each natural enemy are listed in the index and used together in the text at the first mention or major section discussing that natural enemy or both. Scientific names for plants and pests are generally avoided in this book. Plant and pests are more widely known and can be found in many other references. Some crops and pests are mentioned so often that using their scientific names in this book would consume much space and be awkward.

Sources for names used in this book include: *An Annotated Checklist of Woody Ornamental Plants of California, Oregon, & Washington* (McClintock and Leiser 1979), *Common Names of Arachnids* (Breene 1995), *Common Names of Insects & Related Organisms 1989* (Stoetzel 1989), *Fungi on Plants and Plant Products in the United States* (Farr et al. 1989), and *The Jepson Manual: Higher Plants of California* (Hickman 1993).

Index

Major topic discussions are indicated by page numbers in bold type. *Photographs* and major *illustrations* are indicated by page numbers in italic type. **Note** both the scientific name (in italics) and common name of organisms before consulting pages listed in the index. Most text uses only the common name **or** scientific name, not both names. Unless stated otherwise, the names listed in parentheses are the taxonomic family name, except for fungi, where the class name is provided.

proovigenic 56
propargite 45
Prospaltella: See Encarsia (Aphelinidae) 58, 71
protozoa 34
Provado 44
pruning affects natural enemies **11**, 108
Pseudaphycus malinus, Comstock mealybug nymphal-adult endoparasitic wasp (Encyrtidae) 20, 71
Pseudococcus (Pseudococcidae)
  comstocki, Comstock mealybug 20, 71
  fragilis, citrophilus mealybug 20, 71
  longispinus, longtailed mealybug 70, 71
  maritimus, grape mealybug 20
Pseudomonas (Pseudomonadaceae)
  cepacia, fireblight and frost control bacterium 21, 30
  fluorescens, pathogen control bacterium 21, 28, 30
  syringae, decay prevention bacterium 30
psylla, pear 97
Psyllaephagus (Encyrtidae) 58
  pilosus, blue gum psyllid nymphal endoparasitic wasp 21, *70, 70*, 71
psyllid (Psyllidae) 10, **21**, **70–71**, 91
  acacia 21, 70, 71, *91*
  blue gum 21, **70**, *70*
  eugenia 11, 21, **70–71**
  pear 21, 97
  peppertree 21, 70, *71*
  potato 21
Pteromalidae, parasitic wasps **59**, *59*
Pteromalus (Pteromalidae) 59
  puparum, pupal endoparasitic wasp 17
Puccinia chondrillina, rush skeletonweed fungus (Basidiomycetes) 24, 37–38
Pulvinaria (Coccidae)
  delottoi, ice plant scale 23, 64
  psidii, green shield scale 23, 64
Pulvinariella mesembryanthemi, ice plant scale (Coccidae) 23, 64
puncturevine, *Tribulus terrestris* (Zygophyllaceae) 24, 36, *39*
purple scale, *Lepidosaphes beckii* (Diaspididae) 23, 64
Pyemotes ventricosus, grain or itch mite (Pyemotidae) 18
Pyrenone 45
pyrethroids, natural enemy toxicity *11*, *43*, 44–45
pyrethrum 45
pyrgotid flies (Pyrgotidae) **74**, *74*
Pyrrhalta luteola: See Xanthogaleruca luteola, elm leaf beetle (Chrysomelidae) *11*, **12**, 15, 75, 76, 118, *119*
Pythium, damping-off fungi (Oomycetes) 21, 28, 29, 30

Quadraspidiotus (Diaspididae)
  juglandsregiae, walnut scale 23, 64, *90*
  perniciosus, San Jose scale 23, 64
quality, commercial natural enemy *51*, 88

rangeland 38, **39–40**
Raphidia, snakeflies (Raphidiidae) *104*
rearing natural enemies **51–52**
recluse spider, brown, *Loxosceles* (Loxoscelidae) 110, 113
recognizing
  parasitism **56–57**, 75, *75–76*
  predation *79*, *80*
  virus infection **122–124**
redhumped caterpillar, *Schizura concinna* (Notodontidae) 18, *60, 61*
Reduviidae, assassin bugs *41*, **92**, *92, 93*

releasing natural enemies **48-54**, *51*
  green lacewings **100**, *102*, **103**
  greenhouse thrips parasite 72
  lady beetles 84
  mantids not recommended 105
  parasites 60, **65**, **68**, **69**
  predatory mites *50*, **108–110**, *110*
  predatory snails 115
resistance to insecticides
  natural enemies 10, **67**, 109, 110
  pests 10
resmethrin, 45
Rhagionidae, snipe flies **97**, *97*
Rhinocyllus conicus, thistle seedhead weevil (Curculionidae) 36
Rhizobius: See Rhyzobius 22, 23, 49, 85, **85–87**
Rhizoctonia solani, root and crown decay fungi (Agonomycetes) 21, 28, 29, 30
Rhopalosiphum maidis, corn leaf aphid (Aphididae) 14
Rhyacionia frustrana, Nantucket pine tip moth (Tortricidae) 17
Rhyzobius (Coccinellidae)
  =*Lindorus lophanthae*, scale-feeding lady beetle 22, 23, 49, **85–87**
  ventralis =*forestieri*, black lady beetle 85, **85–87**
rice 24, 37, 38
  water weevil, *Lissorhoptrus oryzophilus* (Curculionidae) *121*
rickettsia, *Wolbachia postica* (Wolbachieae) 63
Ringers 44
robber fly (Asilidae) **96**, *96*
Rodolia cardinalis, vedalia beetle (Coccinellidae) 64, 77, 85, 86
root and crown decay fungi 21, 28, **27–31**, *29*
root knot nematode, *Meloidogyne incognita* (Meloidogynidae) 30, *34*
RootShield 30
rose 12
  aphid, *Macrosiphum rosae* (Aphididae) *95*, *121*
rosy apple aphid, *Dysaphis plantaginea* (Aphididae) 14
rotenone 45
rove beetles (Staphylinidae) **81**, *81*, 90
rufous carabids, *Calathus* (Carabidae) 89
Rumina decollata, decollate snail (Achatinidae) 12, 49, 115
rush skeletonweed, *Chondrilla juncea* (Asteraceae) 24, 37–38
Russian thistle or tumbleweed, *Salsola iberica* =*australis* (Chenopodiaceae) 37
ryania 45
rye, perennial 33

sabadilla *43*, 45
sac spiders or twoclawed hunting spiders (Clubionidae) 111, **112**, *112*
Safer's 44
safety, biological control adverse effects **5–6**, *38*, 40
safflower mycorrhizae, *Glomus fasciculatus* (Zygomycetes) 31
Saissetia (Coccidae)
  coffeae, hemispherical scale 23, 64
  oleae, black scale 22, *43*, **65**, *65*
Salsola iberica =*australis*, Russian thistle or tumbleweed (Chenopodiaceae) 37
Salticidae, jumping spiders **113**, *113*
Salvia aethiopis, Mediterranean sage (Lamiaceae) 36
San Jose scale, *Quadraspidiotus perniciosus* (Diaspididae) 23, 64
sand wasps (Sphecidae) **99**, *99*
sap beetle, predaceous, *Cybocephalus californicus* (Nitidulidae) 23, 64, *90*

sawflies (Diprionidae) 22, 118
Scaeva pyrastri, large hover fly (Syrphidae) *95*
scale **22–23**, 64
  black 22, *43*, 65, *65*
  brown soft 22, 64, 65
  California red 8, 22, *43*, 64, 65, **65–66**, *66*, 86
  citricola 22
  cottony cushion 22, 64, 74, 77, *77*, **85**, *85*, **86**, *86*
  dictyospermum 22, 64
  European elm 22
  euonymus 87
  fig 23, 64
  Florida red *66*
  frosted 8, 23
  green shield 23, 64
  hemispherical 23, 64
  ice plant 23, 64
  lady beetles 23, 49, 79, **85–87**, *86*
  latania *65*
  nigra 23, 64
  obscure 23, 64
  oleander 23, *65*, 86
  olive 23, 64
  parasites 49, **64–66**, 77
  purple 23, 64
  San Jose 23, 64
  walnut 23, 64, *90*
  yellow 23, 64
Scaphinotus, snail predators (Carabidae) 24
Scelionidae, parasitic wasps 57, **59**, *59*
Schizura concinna, redhumped caterpillar (Notodontidae) 18, *60, 61*
Sciomyzidae, marsh flies **97**, *97*
Scirtothrips (Thripidae)
  perseae, avocado thrips 106
  citri, citrus thrips 11, 12, 24, *107*, 108, 115
Scolothrips sexmaculatus, sixspotted thrips (Thripidae) 20, *106*
Scotch
  broom, *Cytisus scoparius* (Fabaceae) 36
  broom moth, *Leucoptera spartifoliella* (Lyonetiidae) 36
screens, insect 11, *51*, **52**, *52*
Scutellista caerulea =*cyanea*, black scale ectoparasitic and egg predatory wasp (Pteromalidae) 22, 59
scutellum 59
Scymnus, Homoptera-feeding lady beetles (Coccinellidae) 87
secondary pest outbreaks *4*
seed
  bug (Lygaeidae) 92
  pathogen prevention 30
Seinura, predatory nematodes (Seinuridae) 34
selective insecticides **9–10**, *11*, **12**, **43–46**
semiochemicals 104
Senecio jacobaea, tansy ragwort (Asteraceae) 36
serpentine leafminer: *See* leafminers 18, *72, 72, 73*, 120
sevenspotted lady beetle, *Coccinella septempunctata* (Coccinellidae) 83
Sevin *43*, 44
sheep, *Ovis aries* (Bovidae) 36, 38, 40
shore bugs (Saldidae) 91
silverleaf whitefly, *Bemisia argentifolii* (Aleyrodidae) 25, 68, 69, 103
silverspotted tiger moth, *Lophocampa argentata* (Arctiidae) 18, *73*, *123*
Silybum marianum, milk thistle (Asteraceae) 36
Sinea diadema, spined assassin bug (Reduviidae) *94*
sinuate lady beetle, *Hippodamia sinuata* (Coccinellidae) *83*

Trichogramma (continued)
  bactrae 18
  nubilale 58
  platneri 16, 58
  pretiosum 7, 8, 16, 56, 60
  semifumatum 16, 42
Trichomasthus coeruleus, nymphal-adult
  endoparasitic wasp (Encyrtidae) 22
Trichoplusia ni, cabbage looper (Noctuidae) 8,
  16, 42, 61, 75, 106
Trichopoda pennipes, plant bug adult endoparasitic fly (Tachinidae) 16, 73
Trichosirocalus horridus, thistle-mining weevil
  (Curculionidae) 36
Trioxys (Aphidiidae) 58
  complanatus, aphid endoparasitic wasp 14
  curvicaudus, elm aphid endoparasitic wasp 14
  pallidus, aphid endoparasitic wasp 14, 67, 67
Trioza eugeniae, eugenia psyllid (Psyllidae) 11,
  21, 70-71
Trissolcus (Scelionidae) 59
  basalis, stink bug egg endoparasitic wasp 16,
  60
  euschisti, stink bug egg endoparasitic wasp 16
trochanter, invertebrate 59, 90
Trogoderma sternale, western tussock moth egg
  predatory beetle (Dermestidae) 18
tussock moths, Orgyia (Lymantriidae) 18, 61,
  118, 123
twicestabbed lady beetle, Chilocorus stigma
  =orbus (Coccinellidae) 64, 85, 86, 87
twolined collops, Collops vittatus (Melyridae) 90
twospotted
  lady beetle, Adalia bipunctata (Coccinellidae) 83
  mite, Tetranychus urticae (Tetranychidae) 107,
  110
  stink bug, Perillus bioculatus (Pentatomidae) 94
Typhlodromus: See Metaseiulus occidentalis, western predatory mite 20, 49, 107, 109, 110
Tyria jacobaeae, cinnabar moth (Arctiidae) 36

Ulex europaeus, gorse (Fabaceae) 36
Unaspis euonymi, euonymus scale (Diaspididae)
  87
Uramyia halisidotae, moth larval-pupal endoparasitic fly (Tachinidae) 18, 73
Urophora (Tephritidae)
  affinis, knapweed seedhead gall fly 36
  cardui, thistle-feeding gall fly 36
  quadrifasciata, knapweed seedhead gall fly 36
  sirunaseva, yellow starthistle-feeding gall fly 37
van den Bosch, Robert iii
Vapona 44
variegated cutworm, Peridroma saucia (Noctuidae) 17
variegated leafhopper, Erythroneura variabilis
  (Cicadellidae) 58

vedalia beetle, Rodolia cardinalis (Coccinellidae)
  22, 64, 77, 85, 86
vein, wing 59
velvet ants (Mutillidae) 99, 99
velvetbean caterpillar, Anticarsia gemmatalis
  (Noctuidae) 106
Vendex 44
Venturia 59
Veratrin D 43, 45
Verticillium lecanii, insect control fungus
  (Hyphomycetes) 122
Vespidae, hornets, paper wasps and yellowjackets 98, 99, 99
vetch 33
vineyards or grape 8, 28, 37, 43, 58, 107, 108,
  110, 111, 114
viruses, invertebrate pathogens 18, 116, 122,
  123-124, 124
Voria ruralis, larval endoparasitic fly
  (Tachinidae) 16, 42, 75
Vydate 45

walking sticks (Phasmatidae) 94
walnut 8, 37, 62
  aphid, Chromaphis juglandicola (Aphididae) 14,
  67
  scale, Quadraspidiotus juglandsregiae (Diaspididae) 23, 64, 90
wasp
  digger, mud or sand 99, 99
  parasitic 55-73
  predatory 98-100, 99, 100
  spider 99, 99
  yellow-legged paper 98
water
  boatmen (Corixidae) 91
  striders (Gerridae) 91
waterlettuce, Pistia stratiotes (Araceae) 39
wax moth, Galleria mellonella (Pyralidae) 118
weaver ants, Oecophylla 100
weeds 24-25, 35-40
weevil (Curculionidae) 15, 36-37, 40, 46, 54,
  118, 120
  alfalfa 15, 62-63, 63
  black vine 15, 121
  Egyptian alfalfa 15, 63
  gorse seed 36
  Italian thistle 36
  rice water 121
  Scotch broom 36
  thistle-feeding 36
  yellow starthistle 37, 40
western
  bigeyed bug, Geocoris punctipes (Geocorinae
  subfamily, Lygaeidae) 93
  flower thrips, Frankliniella occidentalis (Thripidae) 9, 20, 71, 106, 109

grapeleaf skeletonizer, Harrisina brillians
  (Zygaenidae) 8, 18, 123
  predatory mite, Metaseiulus occidentalis (Phytoseiidae) 8, 20, 107, 109, 110
  tussock moth, Orgyia vetusta (Lymantriidae) 18
wheast 104
whirligig beetles (Gyrinidae) 81, 81
white
  amur or grass carp, Ctenopharyngodon idella
  (Cyprinidae) 36, 38
  grubs, Cyclocephala (Scarabaeidae) 15, 118,
  120
whitefly (Aleyrodidae) 25, 42
  ash 25, 71
  bayberry 25, 71
  citrus 8, 25, 71
  greenhouse 25, 48, 68, 68, 69, 69
  parasites 11, 25, 57, 59, 68, 69, 71
  pathogens 118
  predators 71, 87-88, 88
  silverleaf 25, 68, 69, 103
  sweetpotato 68
  woolly 25, 71
wing identification characters 59
wireworms (Elateridae) 120
Wolbachia postica (Rickettsiaceae) 63
wolf spiders, Lycosa (Lycosidae) 112, 112, 113
woolly
  apple aphid, Eriosoma lanigerum (Aphididae) 8,
  14, 105
  whitefly, Aleurothrixus floccosus (Aleyrodidae)
  25, 71
World Wide Web information 125-126

Xanthogaleruca =Pyrrhalta luteola, elm leaf
  beetle (Chrysomelidae) 11, 12, 15, 75, 76,
  118, 119
Xylomyges curialis, citrus cutworm (Noctuidae)
  17

yellow
  jackets (Vespidae) 98, 99, 99
  -legged paper wasp, Mischocyttarus flavitarsis
  (Vespidae) 98
  scale, Aonidiella citrina (Diaspididae) 23, 64
  starthistle, Centaurea solstitialis (Asteraceae) 37,
  38, 38
  sticky traps 53, 54, 72
  striped armyworm 16

Zelus renardii, leafhopper assassin bug (Reduvi-
  idae) 41, 93
Zephyr 44
Zeuxidiplosis giardi, Klamathweed gall midge
  (Cecidomyiidae) 36
Zoophthora phytonomi, alfalfa weevil disease 15,
  63